# Horse Speak:
## The Equine-Human Translation Guide

**Conversations with Horses
in Their Language**

## Sharon Wilsie & Gretchen Vogel

Photographs by Rich Neally

TRAFALGAR SQUARE
North Pomfret, Vermont

First published in 2016 by
Trafalgar Square Books
North Pomfret, Vermont 05053



**Disclaimer of Liability**
The authors and publisher shall have neither liability nor responsibility to any person or entity with respect to any loss or damage caused or alleged to be caused directly or indirectly by the information contained in this book. While the book is as accurate as the authors can make it, there may be errors, omissions, and inaccuracies.

Trafalgar Square Books encourages the use of approved safety helmets in all equestrian sports and activities.

**Library of Congress Cataloging-in-Publication Data**

Names: Wilsie, Sharon, author. | Vogel, Gretchen (Equine expert)
Title: Horse speak : the equine-human translation guide / Sharon Wilsie &
   Gretchen Vogel.
Description: North Pomfret, Vermont : Trafalgar Square Books, 2016. |
   Includes index.
Identifiers: LCCN 2016010656 | ISBN 9781570767548
Subjects: LCSH: Horses--Behavior. | Animal communication. | Human-animal
   communication. | Body language.
Classification: LCC SF281 .W55 2016 | DDC 636.1--dc23 LC record available at https://lccn.
loc.gov/2016010656

Photographs by Rich Neally except figs. 2.5 and 8.5 from *Gallop to Freedom* by Frédéric Pignon and Magali Delgado and used by permission from Trafalgar Square Books
Illustrations by Sharon Wilsie
Book design by Lauryl Eddlemon
Cover design by RM Didier
Index by Andrea M. Jones (www.jonesliteraryservice.com)
Typeface: DIN

Printed in China
10

*In loving memory of Zeke and Vati*

*This book is dedicated to the horses that have carried
me past the known and into the unknown. May your
wisdom illuminate the way for all humans who wish to
enter into the landscape of your peaceful world.*

**Also by Sharon Wilsie**

*Horse Speak* (video)
*Horses in Translation*
*Essential Horse Speak*

# CONTENTS



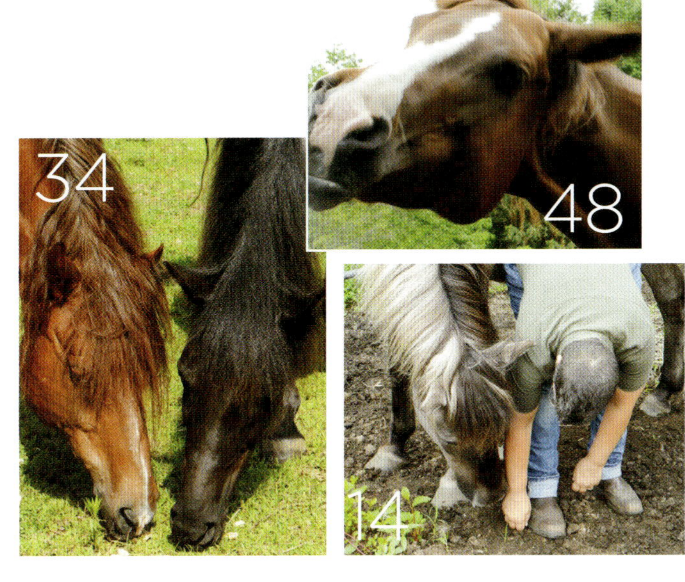

# Everything Means Something

Some of you may remember Mr. Ed, the talking horse, who starred in a black-and-white television comedy from 1958 to 1966. Mr. Ed's lips flapped during the voice-overs, creating a magical effect, especially for a horse-crazy kid like me. I fantasized about what my "someday horse" would say to me—if only he could talk.

Childhood asthma restricted my activities to reading about horses and drawing my herd of model horses. In junior high I was well enough to begin working on a horse farm, and by the end of high school, I was exercising top-level dressage horses. Finally, as a young mother, I was able to have horses of my own. I adopted a large pony, the color and build of the ancient Tarpan horse (an extinct subspecies of wild horse that some attempted to bring back via selective breeding). Rocky was to be the companion for a young, golden snowflake Appaloosa mare named Dakota. I became a student of these two "horse gurus," and in time, even more horses joined my herd. It was as if my childhood models had come to life in order to teach me.

I started to notice the same gestures and movements made by my horses at various times. Whether subtle micromovements or larger postures and positions, I saw each quickly shift into other twitches and motions. My artist's eye was able to freeze-frame the gestures as images. I began to see how the same gesture meant the same thing to *all* horses, and further, how horses were bound to respond similarly, no matter which horse was making that gesture. I suspected I was witnessing *language* and not simply *behavior.*

**Keys** to
Horse Speak:
Introduction
**Leadership (p. 3)**
**Connection (p. 4)**
**Dialogue (p. 4)**

By this time, my adult equestrian career included coaching a collegiate equestrian team, giving lessons and clinics, rehabilitating horses, consulting with horse sanctuaries, and teaching Equine Assisted Learning (EAL). In each setting, I observed the highly ritualized and predictable language of horses. At home, I took notes, made drawings, and watched some more. My passion for understanding the meaning of these individual gestures made me try to mirror them back to the horses. I had the dubious responsibility to come up with clear and consistent movements they might recognize in return. Quite suddenly, I realized my horses were teaching me their kinesthetic (body-language) dialogues. I was being escorted through the basics of language like a young foal learning to be in the herd. And then one day it dawned on me that I really was having Conversations with my horses in *their* language. Learning their language opened a portal into their innate way of being.

I've always admired the work of particular horsemen and horsewomen who seem to be having a private dialogue with the horse. If I was "seeing language" and learning how to use the "words" and "phrases" to converse with horses, perhaps I could help others read, interpret, and act on the subtle cues. I wanted to decode this language, breaking it down into small details anyone could understand. This book is the result of my desire to share what I've learned and help others communicate with their equine partners in the same illuminating and fulfilling ways I can.

Horses must be bilingual. They must speak "horse language" to horses (although if they have been isolated, they may not be effective communicators with other horses). In addition, horses need to speak "human." We use words and inflections to indicate our many levels of feelings, needs, and replies to each other. Humans don't need to rely on subtle visual messages. We express less with our bodies than we do with words. We make so many random movements around our horses that they can be at a loss trying to understand us.

This book, my translation guide to the visual language of the horse, which I call *Horse Speak*, is meant to help both sides of the coin: humans and horses. Just like learning a foreign language, you will first learn distinct movements that are the horse's individual words. You'll learn how to mirror his words visually, in a way he will understand. Eventually, you will be able to combine these gestures in sequences and pose questions to horses. When a horse replies with his own movements, you will understand what he is saying and be able to respond logically. Fluency in Horse Speak comes with the flexibility to have a conversation automatically, calibrated with appropriate intensity. And giving you the means to become fluent is my goal in providing this manual.

## WHAT IS LEADERSHIP, REALLY?

Horse Speak helps you build a relationship with your horse—one based on *trust, respect, unity,* and *acceptance.* Achieving these four elements depends on your ability to express yourself as a leader.

Horses are captive and dependent on us for their very survival. Being a leader actually means letting your horse know that when you show up, all is right and he is *safe.* Does your horse trust you to guide him through the human world? We've taken horses out of their natural environment and asked them to step through doors, jump over obstacles, and get in and out of trailers. Imagine you were in a foreign country, the tour guide

*A true leader behaves like someone you would want to follow.*

didn't speak your language, and he was uncertain and nervous as he asked you to follow him in and out of strange places and spaces. You wouldn't feel safe in his care! You can tell a horse he is safe with you via simple *Conversations* that address his concerns and by knowing how to clearly calibrate *phases of intensity* within the practice of Horse Speak. This will, in-turn, cause you to feel safe with him as you both develop deeper trust in one another.

We all have events that challenge our safety. You might have to bring a horse in during a thunderstorm, or when your horse is sick or injured, practice first aid until your veterinarian arrives. Sometimes, another horse in distress at the barn or horse show can "set yours off." However, the more regularly you have Conversations with your horse in *his* language, the less likely you are to lose your connection with him during a stressful event. This keeps you both safe and *that* is the essence of good leadership.

### The Ideal Connection

The bottom line is, when your horse doesn't feel safe with you, you are not safe with him. Many of my clients are people who have had one "horse romance" in their lives. They tell me about the horse-of-a-lifetime they were effortlessly in sync with—the horse that, even when things were not going well, kept trouble from becoming a disaster. The horse that rebalanced them over the jump, got them through a muddy spot that turned into a bog, stood still when they slipped and fell on an icy patch, and didn't react when a flock of wild turkeys took off from trees overhead. This was the horse that took care of them.

Fortunately, people don't need to go from horse to horse to find this ideal connection. Instead, you can bring resources from within yourself to create a safe and intimate partnership with almost *any* horse. This book is not about training—it is about *transformation*. Yes, most of us once had the blessing of a wonderful horse. Although all horses have much to offer us, the real question is, what do we offer them in return? When we think of being safe around a horse, we always think in terms of *the horse not harming us*. Do we ever consider the horse might be worried that *we might harm him*?

## TRUE DIALOGUE

I offer Horse Speak as a framework for communication. It is not a rigid formula. When you are truly in a dialogue you can never predict how a horse will answer you on any given day. I've now set aside my "normal" horse-training techniques in favor of Conversation with a horse because I can pose a question or pitch an idea to him and see what he thinks about it. Knowing Horse Speak has made it possible for me to know, unequivocally, if a horse is answering my question with curiosity, playfulness, and affection, or just complying with my request out of fear, habit, or training. You'll discover that, as you learn how to have a Conversation with your horse, you become more relaxed and playful. Many of you value your relationship with your horse as much as you value his performance. Deeper bonds of friendship will blossom as you show your horse you are willing to listen and learn his language instead of just expecting him to respond to yours.

What ultimately separates us from animals are our own misconceptions. Traditionally, we've held that groupings of behavior are instinctive and outside the animals' control. Their movements have been labeled as "surface" behaviors. This definition is still widely held by many trainers and is a limited view that belies horses' acute sense of self. We believe, since animals don't think in our words or have a sound-based language, that they have no internal dialogue. Yet, horses use their language to link to the internal narratives of other horses all the time. They assist each other and change each other's minds. Horse Speak now allows them to link with us. It is the way our horses will ultimately help us begin to think differently about them.

### Species Relationships

Humans do not have to be acutely aware of their environment in order to survive. We don't have to pay attention to detail. Our words are more important than body language so we don't even have to pay that much visual attention to each other in Conversation. Horses' language, on the other hand, is rooted in their worldview as prey. They've survived in spite of being the food of choice for a variety of large carnivores. They possess an intense awareness of their environment with almost

360-degree vision, acute hearing with big upright ears, a great sense of smell, and the ability to run fast. They see your facial expression, body language, clothes, and hair, and every twitch you make. Your intention to move in a certain direction is easily read well before you actually do it.

Humans and horses are both social beings. We have this in common. When you are with a horse, the horse considers you a part of his herd, even if it is only a herd of two. When a chicken is in the pasture, the chicken is now counted in the herd. All of us have seen cats, goats, and dogs in a herd with horses. Horses offer us this natural inclusiveness, as well. It is one reason we may feel so good when we are with them. But it is also a reason for becoming more in tune to what our body language may or may not be saying to them.

There is an unfortunate idea in the horse world that we cannot let the horse "win" or resist our requests. The truth is, most horses are designed by nature to accept a submissive role; very few are truly "alpha" animals. Learning their language brings new understanding of their code of honor and how it can be part of a successful relationship with horses, no matter what. Perhaps the Conversations will bring you new insights and show you secret parts of yourself, as well. When we allow a horse to maintain his dignity, while utilizing his own code of ethics to engender mutual respect, everybody wins. Allow yourself to evolve as your relationship with your horse transforms: investing in this is taking steps on a journey to becoming a whole, healthy, and balanced human being.

My deepest hope is that when people understand the language of the horse, it will become impossible to abuse a horse "by accident" because someone didn't know what the horse was saying. We can balance our expectations with our horse's preferences, and at the very least, hear him out, without being afraid that giving him a say in his life will result in him being spoiled or ruined.

Some horses have more opinions than others. By practicing Conversation, you create trust, respect, unity, and acceptance. I hope I've contributed some tools that humans will use peacefully and joyously with horses. The warmth of a connected friendship is all the reward any of us need for the small amount of time it takes to learn and implement this mutual language.

## 12 EASY STEPS

In clinics and with students, I teach Horse Speak in 12 Easy Steps, and so that is how I have organized the chapters ahead. It is my hope that a sensible progression such as the one laid out here will make learning this means of communication both appealing and infinitely achievable.

*Horses are sub-missive by nature; they deserve to keep their dignity.*

Horse Speak is acquired in layers, and it is necessary to learn some skills in a general sense early on, only to return to them later and explore them in depth. And so I begin with fundamentals that are important to understand and be able to employ every step along the way. Two areas are of particular importance to this work: breath and body language. You will find pages that detail my understanding of the meanings behind equine communication via breathing and physical gesture, before we progress to what I call the Four Gs of Horse Speak: Greeting, Going Somewhere, Grooming, and Gone. The Four Gs are the framework of equine interrelationships, and almost all "discussion" between horses—and I feel, between horses and humans—can be categorized under one of these headings.

While I can help define and describe different modes of "listening to" and "talking to" your horse, as with human speech, much depends on that which is anything but obvious. With Horse Speak, it is the level of intensity, or pressure, that accompanies a gesture, movement, or stance—no matter how subtle, the slightest motion can in fact deal great power. Any individual truly committed to being able to communicate effectively and fairly with horses must practice regulating intensity, and I offer a number of exercises to help you begin to manage what's going on inside your body, as well as what is going on outside your body.

In addition, get ready to sample dozens of what I call "Conversations," distributed throughout all 12 Easy Steps. The Conversations provide step-by-step templates for you to follow in your day-to-day interactions with your horse: how to read and "hear" what your horse is saying, and what you should do and "say" in reply to end in a happy place.

Finally, you'll meet Joe. He is a fictional horse with the kinds of issues common to many equines today. You can follow along as I use Horse Speak to work with Joe and his owners, from our first meeting to the resolution of many of his problems. The details I can provide in describing the process of interacting with such a horse supply an invaluable glimpse behind the scenes, tools you can most certainly use, and cause-and-effect scenarios molded with one thing—educating the reader—in mind.

Horse Speak is not a training method, nor does it specifically teach you how to be a better rider. However, whatever discipline you prefer, whether you subscribe to classical horsemanship or natural horsemanship or something else altogether, Horse Speak can complement your methods and improve your relationship with your horse. Just wait until you see what it will do for you.

# Building Your Foundation

**W**here do we begin when learning a new language? Let's start with zero and go from there.

## FINDING YOUR "INNER ZERO"

Horses' survival depends on their ability to observe what is happening around them and see each other's subtle messages. This is only possible when the "herd" (same species or multispecies, of many or of only two, in the barn or loose in the pasture) is calm (fig. 1.1). When one horse starts to pick on another, the others in the herd may quietly separate the two, adjusting their positions until there is a state of calm again. In conflict, opponents are likely to kick or bite at one another; perhaps one will drive the other away. Then, as quickly as the disagreement began, it ends. Horses do not thrive on stress and excitement.

I regularly observe my herd and usually see some standing a few feet apart, some lying down, and others grazing—but all striving for a steady state of calm.

*1.1 Horses prefer a calm and thoughtful state of being, whether at rest or on the move, as here.*

## Keys to Horse Speak: Step 1

**Inner and Outer Zero (p. 8)**
**Observation (p. 12)**
**Mirroring (p. 18)**
**Sense of Humor (p. 21)**
**Good Questions (p. 22)**
**Pausing (p. 23)**
**Breath Messages (p. 24)**
**Bubbles of Personal Space (p. 28)**
**13 Horse Speak Buttons (p. 36)**

Many of us may try to be calm around horses as a general rule, but for the process of learning Horse Speak, it is essential. Being "quiet inside" allowed me to be more "present" (in the moment), which is how I was able to observe the subtleties of equine language to begin with. Establishing ways to be "present" with your horse is how *you* will see their language, as well. I call the state of being present in the moment, being aware and calm, *Inner Zero* (fig. 1.2).

Zero is the first thing I teach any of my students. I ask them to imagine a favorite peaceful spot—a happy place. Some think of a favorite song, a lovely image, or an emotionally fulfilling memory. We all have individual gateways to our Zero. Practice the feeling that comes over you when you think of the trigger image, sound, or memory and return to it repeatedly. This is Zero on the *inside*. Sometimes I have my students stand in the vicinity of their horses, balance their weight on both feet, and breathe into their bellies. Then we all notice if and how the horses respond.

It took me years to realize how important my own state of Zero was for successful Conversations. I've developed another tool to help me stay calm and present no matter what the horse is doing. I can stay in my Inner Zero by thinking or saying out loud the words, "How curious." When you say, "How curious," it neutralizes emotions, no matter what the horse is doing. Or you may want to try the word, "Interesting," which is also a nonjudgmental expression. No matter if things are going well or badly with your horse, find your own trigger words or phrases that will help you stay calm and present.

## MANAGING "OUTER ZERO" AND ADJUSTING VOLUME

The best way to clearly communicate with horses is with *planned, deliberate gestures* (fig. 1.3). When your inner intensity level is calm or Zero, it does not mean being *too* quiet or tentative. Horses prefer us to move with purpose, be aware of the environment, and be confident leaders. Quiet, confident assertiveness equates to "calm" for horses. It tells them, "All is well." In addition, being calm promotes mutual trust because horses aspire to stay in this state.

Most of the time horses *whisper* with their body language. A peaceful head nod, tiny twitch of the tail, the intention to pick up a foot with a shift of balance, these say all they need to another horse (fig. 1.4). Occasionally, a horse shouts or

*1.3* Movement that has purpose communicates clear intention.

yells a message at another with extreme gestures or movement, but usually horses say what they need to at the lowest volume necessary. Even if they "turn up" the volume, they are experts at then turning it back down, calibrating the movements of their language precisely.

Horses do not want to waste energy by bickering with another all day or holding on to a grudge. In order to ensure safety in the wild, they want to stay quiet and not call the attention of predators with noise or hoofbeat vibrations through the ground. When a horse asks another for space, he may simply swish his tail, then throw his head, stomp a foot, and finally, kick or bite. As soon as the other horse moves, everything goes back to Zero on the *outside* as well as on the *inside*.

Horses are good at this, but we have to practice. For the sake of mutual Conversation, it is vital you understand how to adjust the volume of your own movement in response to theirs. You can learn to mimic their language and adjust your volume with the appropriate intensity.

To make this easier, I have assigned numbers to represent the levels of intensity of physical movement or volume you might use in a Conversation with a horse. I use five numbers: *Zero, One, Two, Three*, and *Four*. As the numbers increase, so does the size and/or emphasis of your movement. I like to use numbers to label these "levels" of Conversation because there is no emotional attachment to them. Indeed, there should never be any negative emotion in any of the Conversations you have with a horse, no matter how large your gestures become.

In general, Zero on the *outside* is like Zero on the *inside*: it means complete calm in mind and body (figs. 1.5 A & B). What does Outer Zero look like? Zero intensity, energy, or attitude doesn't mean you are not doing anything; Outer Zero is the *way* you are doing it, physically, and has different postures. You could:

- Bend one knee to cock your own hip sideways—the way horses do.
- Put your hands in your pockets, soften your gaze, and drop your head slightly while breathing deeply.
- Look at the ground and "blow out" a sigh.
- Make your body look limp like a rag doll.

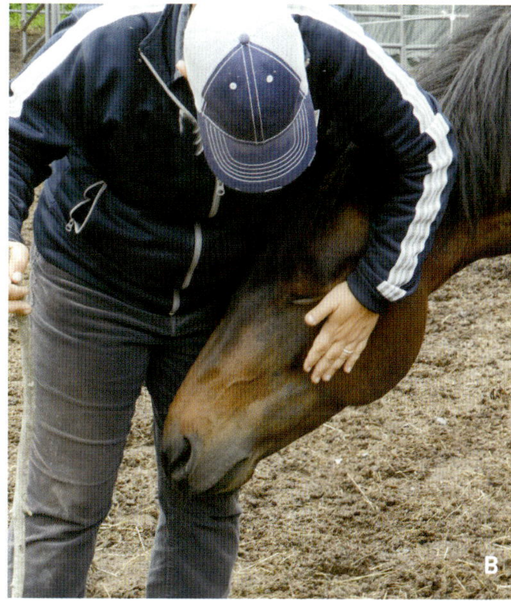

***1.5 A & B*** *Inner and Outer Zero (A). Being Zero inside and out invites more pleasurable moments with your horse (B).*

We all develop our own versions of Outer Zero around our horses.

The next phase of intensity is Level One, which is *intention* (your determination to act a certain way) alone without much movement at all. Level Two volume adds *motion* to intention. Level Three adds *movement toward the horse* and may include *touch*. Level Four volume is most intense including the largest of *gestures* in any given situation. Another way to look at this is to consider the five phases of physical movement or volume as: Calm, Thinking, Asking, Telling, and Insisting. Remember this is *not* about training—just about language. Different intensities of movement are the *adjectives* in Horse Speak. (We examine them in more detail, beginning on p. 100.)

Your goal is to habitually return to Outer Zero no matter what level of intensity has been used for any exercise or Conversation. But *ceasing* intention, motion, gesture, or touch ("turning down" the volume) is the hardest thing for any of us to learn. If you practiced nothing but returning to Inner and Outer Zero whenever and wherever it might be necessary, many of your problems with your horse would melt away. Horses let go of stress and effort more easily than humans do. We tend to relive an emotional charge long after the incident is over. Zero can help change this.

*1.6* *The horse uses his entire body to communicate his thoughts and feelings.*

*1.7* *Horses are always watching you when you are with them.*

## OBSERVATION

Horses communicate by using their entire body. Humans call this "body language." As already mentioned, we see this kinesthetic language both in the smallest of twitches as well as in grand movements, postures, and positions. Some of the body language I'll discuss in the pages ahead includes an amazing variety of breaths, facial-expressions, tail swishes, and angles at which a horse holds himself. Everything horses do, even when standing still, means something to other horses (fig. 1.6).

Becoming aware of Horse Speak is pretty simple, really. You just need to pay attention whenever you are with your horse. Since horse language is predominately visual, horses watch each other closely even when it doesn't seem like they are doing much of anything. Horses are always watching you closely, as well, when you are with them (fig. 1.7). So plan to become like horses when you watch them. Simple observation makes your horse more interested in you because you are showing him you are interested in *him*. Consistent observation is the second step you'll take (after learning to find and manage Inner and Outer Zero) to transforming your relationship with your horse.

Horses communicate with us all the time in ways so subtle we don't notice much of it. For example, you may throw hay out at feeding time, then head back to the house to get ready for work. How many of you glance back and notice that one ear of every horse is still on you, even though they are eating and you are walking away? What is so important about a tiny thing like this? It is the kind of thing horses do with each other. *All* movements in a herd are noticed. Over time and with practice you'll learn to notice every little thing, too.

*1.8* *When you start observing your horse, he will take notice!*

Just mucking out stalls? Observe your horse as closely as you can as you work around him. Turning your horse out with buddies? Watch to see when and how the horses interact with each other. Gazing out the window at the herd in the pasture? Notice what they are doing, how they are moving. As you learn more about their language, you'll realize a lot is going on even if you never thought so before (fig. 1.8). Observing your horse is an "approach" message: he will know you are watching him, trying to see and understand him, and as a result, he will become more engaged and trust will grow without you even realizing it. One day, your head-shy horse will bring his muzzle closer to you for a breath exchange or he will do a huge head bob as you come toward the gate. Your attentive observation is reaching out to him in a new way—and it is one he understands. You are now *listening* to his body language.

*1.9 A–C* Mother Nature gave horses a sense of humor (A) and curiosity (B), which we can often see in their expressions. Their desire to be playful is a lot of fun when we can tap into it (C).

When we are not paying attention with regular observation, a horse has to resort to the equine equivalent of "shouting" at us, which means he communicates his happiness, fear, confusion, pain, and ideas in grand gestures. Some of these big movements may even be called vices because we don't know why he is doing them. For example, he fidgets on the cross-ties, steps on your foot, or rushes out the stall door. We might label him as stubborn or stupid, but *we* are the ones being callous. We just haven't noticed the more subtle ways he has been trying to tell us about his concerns.

Some horses shut down and do not even attempt to show us what they think or feel. They have learned there is no point in making the effort. Why try to communicate with a species that doesn't take an interest in learning, seeing, or acknowledging their perceptions?

Half a Conversation is listening and the other half is talking. If you aspire to "see" your horse's language, you must remember to listen *even when the horse is saying things you don't want to hear.* When your horse gets emotional and you do not know what to do, keep yourself safe, then go to your Inner Zero. At least this will afford you the right state of mind to observe your horse and perhaps come up with a helpful strategy. You will see your horse with a whole new level of awareness once you know the words he uses.

## MAPPING THE LANGUAGE

I'm really grateful that horses' language is visual, as that makes it easier to "map." I think of each individual gesture or posture as one "word." Horses can combine these "words" to express complex ideas—what we would think of as "sentences." For example, when your horse twitches his lips as you approach, he is saying, "I'm having a good day." When you wriggle your lips back, you say to him, "I'm having a good day, too." In Horse Speak we use another phrase called *Sentry Breath* to describe your horse snorting or blowing at an object (see more on p. 27). And, when you blow in the same direction as if you are blowing out candles on a birthday cake, looking at the ground and pretending to chew gum, you are telling the horse you've seen the "bogeyman," blown him away, and all is well.

## Signs of Emotion

Dogs, cats, and people display emotion in a *direct* way. Cats and dogs come up to you when you come home from work, follow you around and wrap themselves around you. Horses, however, being prey animals, display emotion *indirectly*—they express affection, for example, by giving you space and by paying attention to you. They express love by treating you as they treat their horse friends: by being calm, bonding, letting you into their space, and sometimes initiating touch in the way they do with other horses.

We enjoy the way we feel around horses. We can feel enveloped in the state of peace and calm, which is the horse's Inner Zero. We ride to merge with this body that is so much larger than ours. We want to feel that our heart is at one with his enormous heart. I believe we all still have this need. It is deeper than a learning technique, deeper than an award-winning performance, deeper even than having fun. It is the need to communicate love and affection to your horse and to know— without a doubt—your horse offers love and affection back to you.

Mother Nature endowed horses with a sense of humor, curiosity, and a huge "play drive" as if to compensate them for having to worry all the time about being eaten. These are all expressions most horse owners can recognize—for example, when a horse wiggles his ears, he is demonstrating his sense of humor. Playfulness and affection are expressed sometimes via sideways ears or big huffing breaths (figs. 1.9 A–C).

*1.10 The more defensive the message, the higher the horse will hold his head.*

Then, there are common defensive messages: For instance, your horse might say, "No!" with a tail swish and a hind foot stomp. A horse holds his head up high when he perceives a threat—it is less likely that the lion or wolf will grab his muzzle when it is held high (fig. 1.10). A high head can also indicate the horse is confused, but when truly scared, he lifts his head and pulls it backward out of what he perceives to be harm's way.

You may have seen a horse running at full speed with grass or hay hanging out of his mouth. The

*1.11* A startled horse cannot swallow when the soft palate closes as is its function, allowing him to breathe through his nose.

horse has a soft palate that closes over his esophagus, thus allowing him to breathe through his nose but not swallow (fig. 1.11). Due to the length and position of the horse's soft palate, he is one of the few animals that can only breathe through his nose. I always notice how the horse's eyes seem to bulge when he is startled, as if enabling him to better see into the distance.

Adrenaline makes it possible for a horse to run for a short distance at high speed. When he is panicked, his shoulder blades, which are not attached to the skeleton, rotate upward, literally making him taller. This releases the back of the rib cage so he can take bigger breaths.

One winter evening I was at the gate, ready to bring the horses in for the night. Instead of coming to me, they trotted way up into the field and looked at something in the woods. So I looked, mirroring what they were doing but not seeing much. Then my eye caught some movement low along the ground. It was a large gray owl. We all watched until it drew up and landed in a tree. I could have been annoyed by their antics while I was standing at the gate—it was cold and time to bring them in—but instead I chose to do what they were doing and slipstreamed into their world for a moment.

Another day, this time in summer, I went out to weed the garden in a ball cap instead of my usual straw hat. My mare came over and was very obviously looking at the cap. I don't really know why—maybe I reminded her of someone else or maybe she was noticing a different detail in her environment (a detail that happened to be on my head). She looked at the cap with concentration for a few seconds and then, as if catching herself being rude, looked off to one side. So, of course, I looked to that side, too. I wanted her to know I saw the gesture. Then she dropped her head, licked, and chewed. I did too. It was a sweet, still moment in my day.

You might also know a horse that pushes into you with his forehand. Pushing you means he is claiming your space; he considers himself above you in herd hierarchy. A pretty extreme version of this defensive message one horse can do with another is a sideways body slam—it is almost like a martial arts move.

These are just a few examples of the kinds of Horse Speak you'll become familiar with in the pages ahead.

## MIRRORING

As you will find, horses have a highly ritualized and predictable language. It is expressed through specific body language. When you understand how and what they are communicating you can mimic their body language back to them. It's called "mirroring" and it makes Conversation possible.

*1.12 Mirroring your horse is fun and teaches you a great deal.*

As children, many of us pretended to be horses. Remember how you "trotted" and "galloped" around on your two legs, tossing your mane, and jumping over things? Obviously, horses are quadrupeds and have a horizontal orientation while we are bipeds and more vertically inclined, but we both still have heads, faces, necks, torsos, and legs. There is a difference in the way we express movement or gesture, but nonetheless, our imitation of horses' language is always rooted in the way they visually communicate with each other (fig. 1.12).

Mirroring a horse is a great opportunity to be a student of nature. Simply look where he looks and stop when he stops, as if you are playing "Simon Says." As silly as it might feel, do some freestyle mirroring of your horse. Without any goal in mind, try to translate his body language into some movement of your own. This may be challenging because we've been taught we always have to be in charge around horses. In fact, perhaps the only Conversation you've ever had with a horse is, "I'm the human and you're going to do only what I tell you." Freestyle mirroring may free you of some ingrained conditioning and bring you to an authentic moment within yourself (figs. 1.13 A & B).

Observation and mirroring give you the space to read a horse's individual personality. It is a great relief to let go of our human ways and simply be with a horse in his sense of time and his sense of relationships, emotions, and thought-processes. Letting go of our human agenda and ego, even for brief periods as we practice Conversation, fosters deeper connections with our fellow creatures.

*1.13 A & B* *In these two photos I am mirroring the horse's head (A), and then she is mirroring mine (B).*

## THE ART OF CONVERSATION

Once you begin to see his language you will also start to see how the horse sometimes has ideas when he is with you. When you have established communication through Horse Speak, you can also see how interested he is in *your* ideas. Most of a horse's ideas are about relationships with other horses, the negotiation of personal space, and the comfort of food and water, of course. Breeding horses think about

cycles, birthing, and the raising of foals. It wouldn't naturally occur to most horses that it might be fun to push a big ball around, jump a course of obstacles, or "dance" with a two-legged human beside them or on board (fig. 1.14).

Horses love to be praised. They take pride in their accomplishments with us and enjoy being admired. When they feel "listened to," many horses will "claim" their performance and enjoy showing off.

All horse training uses a version of shaping behavior in repetitive exercise. Schooling often involves desired movements being rewarded and unwanted movements being punished. When a trainer is consistent, there will be some successful behavior modification. I know because I've been a trainer (fig. 1.15). Horses are generally trained to perform *without* thinking. They don't understand why they are being forced to complete the same patterns over and over. They are not invested, present, or even thinking about what is happening. They are not having *ideas.* Now, the moment I see a horse is *thinking* about what I am asking for, I *stop asking.* This "release" rewards him for *thinking.* This makes it easy for him to be successful and impossible for him to get it wrong. This also relaxes the horse and builds his confidence. The idea of Horse Speak is to assist both you and your horse to move past limiting beliefs or old experiences and move into a new understanding of each other, where you both have ideas worth exploring.

## PRAISE
### IN PRACTICE

Once I was riding a highly trained Warmblood. He had been pushed into performing his dressage movements, and at different times, he would brace and seemed to wait for me to discipline him. I wanted instead to suggest ideas for movement and see if he wanted to do these things because they would be fun for us both. When he felt graceful, I praised him. Then he would forget we were having fun and brace again. I told him out loud, "You can brace all you want and I won't hit you, but didn't it feel better when we were flowing together?" Eventually, he began to offer more and more beautiful movement as he blossomed with my positive attitude.

1.15 *A thinking horse may put his ears backward in concentration. This is similar to a person grimacing while taking a test.*

## A Sense of Humor

I enjoy being with horses. I smile and laugh a lot. Horses seem to appreciate my good mood. If you have a sense of humor, learning Horse Speak will be funny at times. Horses' body gestures speak volumes to other horses. But our body language is gibberish and random for the most part. So, when you start to use gestures to mimic theirs, you have to chuckle a lot. Laughing at yourself helps keep you from getting frustrated because horses may not catch on to what you are doing right away. Having a sense of humor as you learn will help you stay relaxed and open. It will also keep you from becoming too goal-oriented.

You do not need to be super-coordinated to have a Conversation with a horse.

The horse has evolved to have great tolerance for humans. You may just be guessing what to say back to him with your body, but keep practicing. Horses always seem amused when we are trying to learn their language, no matter how clumsy we are at first.

In the horse world there is a tendency to be very serious-minded. Why are we so dour with horses? Even if love of and joy with horses is something we do quietly and experience inside, can't we at least let our horses know we enjoy them? Conversation with a sense of humor is a way to not only make friends with the horse, but it can be rewarding as the horse offers affection and warmth to a depth I almost guarantee you've never felt before. Let yourself be less tense, less driven, as you notice the little things your horse does. Release your "agenda" and talk out loud with your horse. Let yourself smile, laugh, and be loved.

Using your Inner and Outer Zero, "How curious" attitude, and a sense of humor will a go long way toward bridging the gap between your two worlds.

## Ask Good Questions

You initiate Conversation with a horse by posing a question. No two horses will answer you in the same way. They have distinct personalities and respond to you based on their needs or preferences. It is up to you to observe and interpret their answers in response to what they communicate. Your next movement or gesture should be an appropriate reply. You then watch for their next move—and so on—until the Conversation is complete. Once you know Horse Speak you will see how your horse is asking you questions all the time and watching and waiting for your answer.

Your own horses are accustomed to your routine actions, so at first it may take several repetitions of your questions before they are even curious about what you are doing. It will seem as if they are waiting to see if you will keep up your efforts. In addition, although the gestures take seconds for you to do, you may not see the horse's side of the Conversation right away. Remember: Every twitch, lean, ear flicker, or look is an *answer* to your question. So it may be that your horse is responding to you but your eye is not yet trained to see it. His response may be too subtle for you to read. Given time and practice in observation as we discussed on p. 12, you will start to see the small gestures. You will be amazed how much your horse actually says when it doesn't look like he is doing much at all.

Sometimes we get an answer we don't want to see. Say your horse says, "No" with a swish of his tail and a hind foot stomp. (Read more about this on p. 127.) You may find a way to take a different direction, one that you both are comfortable with.

You can be sure he is seeing your language attempts when he starts to ask you questions in return. You may even miss these at first.

## Pausing

*Pausing*—going momentarily to Inner and Outer Zero (see p. 8)—is one of the most important parts of any Conversation you have with a horse. I discovered that the pivotal point in a Conversation is not while posing a question or pitching an idea, not when the horse shows me his answer, and not even when I respond in dialogue with him. The pivotal point in the Conversation is the *interval in between actions*: the Pause—the quiet moment I give a horse so he has the space to think and consider. For example, when a horse answers me with a twitch

## ENGAGEMENT
### IN PRACTICE

At a recent clinic, there was a Mustang that seemed happy when he recognized my Horse Speak language skills. He was one of two horses loose in an indoor arena, and I asked all the attendees to try a series of simple Conversations with him and his buddy. If one of my students forgot part of the Conversation, the mustang would cue the student, demonstrating the expected gesture. The Mustang even prompted the other horse when it seemed he didn't understand the Conversation. It was a wonderful example of a horse deliberately engaging humans in Horse Speak.

*1.16 Learning when to Pause, return to Zero, and give the horse time to think, is crucial to the art of Conversation.*

or a flicker of his ear, I stop (Pause) to let him know I see his answer. Horses may require longer Pauses as they consider an answer to give you. When they are truly thinking, it can be at a snail's pace.

Even if you don't understand the meaning of a horse's movement or gesture, stop all cues and go to Outer Zero. Pause and give yourself time to think (as well as the horse) and decide what your response or next question will be. Your horse may just be taking a moment to find a really good answer for you, as well.

Just like us, some horses need more time to think than others. At times, a horse may seem to tune us out in the Conversation. Realize he might be thinking things through and not ignoring you. Thoughtful horses are worth waiting for because when these horses answer, it will be clear and dramatic. Horses can transform in Conversation, releasing confusion, showing us in Horse Speak how much is shifting internally for them.

Hitting my own "Pause button" has become somewhat automatic. But I know from the lessons I teach in Conversation, coaching my students when to *stop* cueing or asking when to *Pause* can be the hardest thing to learn (fig. 1.16).

### No Wrong Answers

You may get it in your head a cue or question from you is going to cause the horse to reply in a certain way. Well, guess what? When your horse flicked his tail at you, it was his answer. Anything a horse does can be your answer when you know Horse Speak. When you are truly *in* a Conversation, you do not anticipate a certain response with a horse any more than you would in regular Conversation with a person.

In any true Conversation there is no wrong answer. I pose a question or pitch an idea and ask the horse what he thinks about it. I grant a horse his inherent right to give me an answer of his own, not one defined by his training. By letting my horse be authentic, I just may learn a lot about myself in the process.

## BREATH MESSAGES

Most of us do not use breath as a communication tool, but when it comes to having Conversations with your horse, breath messages may be the best way to show him things are changing for you both. If you practice deep breathing, for example, each time you approach your horse, in time he'll believe and see you are trying your best to be aware of what is a foundational tool of Horse Speak.

## Greeting Breath

A horse's lungs extend forward and backward from his shoulders within his body. He communicates from the depths of his body with a variety of types of breaths. One is the *Greeting Breath*. When he huffs three breaths toward another horse from a distance, you may notice your horse uses an extended out-breath with the last huff (fig. 1.17). Horses also just sniff at each other as a "drive-by" greeting.

Try blowing three huffing breaths at your horse from a visible distance. Watch for his nostrils to flare as he greets you in return.

## Beckoning Breath

A horse can use a *Beckoning Breath* to welcome another horse or a human. You will only see a subtle nostril gesture that accompanies this deliberate breathing: the nostrils widen and the horse breathes softly. Your horse "breathes you in," inviting you into his space. It is such a sweet experience once you notice this affectionate gesture.

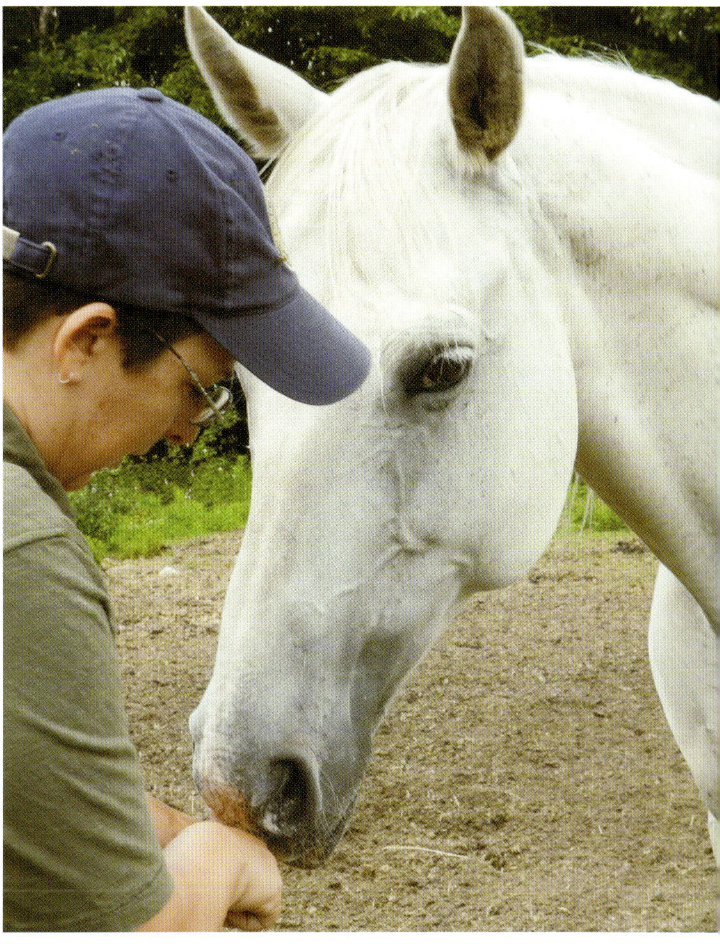

*1.17 When Vati greets me with her breath, I blow gently at her in response. The photos of Vati in this book show the very first time she and I interacted using Horse Speak.*

## Interested Breath

A horse uses short *intakes* of breath to indicate interest, as if he is taking whiffs of a pleasant scent. You can indicate deep interest in your horse with three inward whiffs (as opposed to outward "huffs" as in a Greeting Breath—see above). This can move from mere interest to friendship when a horse sniffs another's neck—a wonderful practice for you to mirror or imitate, especially before any activity, such as grooming or mounting, or simply while visiting in the stall or paddock with him.

## Nurturing Breath

Another important breath message sounds like an *inward* sniff or snort. Mares use this breath with newborns to urge them to stand and nurse. I call this the *Nurturing Breath*. Mimic this breath by clearing your throat with an inhale—it can resemble the soft snort a pig might make. I once caught an injured horse no one else could get near using the Nurturing Breath. He looked at me as if to say, "Really, you care that much?" I was able to walk right up to him. This inward snort is a great tool to keep in mind; you never know when you will need to distract or soothe a horse in a critical situation.

## Relaxing Breath

A long, soft, *blowing out* of breath is used by horses that want to relax one another. This is the same *Relaxing Breath* you might use when blowing softly toward the horse. Horses exchange breath all the time, and as they do they will usually softly clear their noses, sounding a little like a dolphin's breathing through a blowhole. Three short, gentle, whiffing breaths out near the muzzle mimics another "friendship greeting" recognized by all horses.

## Yawns, Big Sigh, and Shuddering Breath

Horses use breath for emotional release. They "blow off" an experience by clearing their noses and shaking their heads. It is as if they are "clearing the air" before moving onto the next thing. Horses *Yawn*, as we do, when we are waking up or napping, but Yawning in horses also has an important message that they are "letting go."

*1.18* Horses demonstrate a "release" through a number of breaths, sometimes paired with a shake of the head or body.

The *Big Sigh* is what you hear when you blanket a horse on a cold day, place a carrot in the dinner tub, or maybe even when you dismount. The Big Sigh is *agreement*. It could translate into the words, "Yes!" or "Amen!" Horses sigh with each other all the time. When you see two horses grooming, for example, at the end they may both release a Big Sigh. My whole herd takes turns giving me Big Sighs when I feed them their last hay for the night.

Another breath of release is when a horse draws two short half-breaths in and then exhales in a long out-breath. I call this the *Shuddering Breath* because it is like the big breath in that we take at the end of a good cry, with a few "hitching-in" breaths before we let out a Big Sigh. Anyone who is tense or anxious should try several of these breaths before mounting. It is the perfect breath to relax and clear the air for both you and your horse.

These types of breath say, "I am letting this go." I am greatly rewarded when the horse shows me he is releasing tension with this breath and gesture (fig. 1.18).

## Sentry Breath

A short forceful snort out is connected to fear. A horse will blow or snort at something in the environment that is of concern. I call this the *Sentry Breath*. You can calm the horse by mirroring this breath: hold your head up, look in the direction he is looking, and blow as if you are blowing out many birthday candles on a cake. I have had great success with panicky horses using the Sentry Breath. The trick is to do it consistently until it becomes a habit. As with other forms of Horse Speak, you will learn you are finding ways to say  to your horse, "I see that bogeyman, too" (photos 1.19 A & B).

## Trumpeting

Trumpeting is rare but can be heard when penmates are separated from each other or when an unknown object causes fear and remains a factor (a perceived threat over a longer period of time, as opposed to the brief snort of warning in the Sentry Breath—see above).

*1.19 A & B  In this photo we see Luna using Sentry Breath as something catches her attention (A). Only moments later she gives the "All's clear" message (B).*

It sounds a bit like an elephant vocalization. I once heard my horses trumpeting—there was a coyote on the other side of the stone wall in no hurry to move along.

### Deliberate Breathing

A horse that is relaxed and breathing deeply may look pregnant because of the distention of his belly. By comparison, a horse that is not breathing normally is also holding tight in his underbelly and will look "tucked up." When worried, horses may minimize their breath, just like we do. You can help *any* horse relax to a greater degree when you take big breaths around him. I sometimes just lean my head against the side of the horse at his girth line and take big breaths until he joins me with a great Big Sigh. As you'll see in the steps ahead, Deliberate Breathing is a key component of our Conversations with horses.

## THE BUBBLE OF PERSONAL SPACE: CIRCLES AND ARCS

In order for a herd to move safely at high speeds, horses must practice negotiating spatial relationships with all the others in the group. These negotiations include determining who is the leader and who should follow whom when the herd must run from a predator.

*1.20 Horses are very aware of their Bubbles of Personal Space. We need to pay attention to their space just as we expect them to pay attention to ours.*

The *Bubble of Personal Space* around a horse is circular, the same as ours (fig. 1.20). In fact, when a horse is calm and has time to think, he will instinctively move in predictable circles or arcs (part of a circle). Fearful or panicked horses run in straight lines. Normally, the size of the circles or arcs changes. When a horse yields his space to you by moving his head over and taking his feet with him, he is actually yielding his place along his circle or arc. Even if a horse seems to be walking straight toward the water trough in the field, watch carefully and you will see subtle shifts in the forehand to his right or to his left. And if you watch horses move in a group with each other, you will see them walk around each other in arcs, tracing the borders of their circular Bubbles of Personal Space (fig. 1.21)

*1.21* The horse moves in predictable circles and arcs related to his Bubble of Personal Space.

# Circles and Arcs
## IN PRACTICE

One day, I was looking out my kitchen window and noticed something odd happening in the field. My old gelding was shooing away his companion mare, and something about him didn't look right. Each time the mare came in close, he made a grimacing face and sent her trotting away. She'd circle him again at the trot and return to check in. I watched for five or six circuits—it actually looked as if my gelding was longeing the mare! When I saw the gelding paw the ground, I could then tell that was a sign of pain and was sure he needed my help—likely a colic episode. But it was the mare's circular dance that caught my attention and ensured I noticed his distress early enough to keep it from progressing.

Although most of the time horses only move in half-circles (arcs) around each other I saw this full-circle phenomenon another time, as well. I had two horses at liberty in a clinic and was *Sending* one (I'll explain this on page 33) with a *Driving message*. The horse would respond by moving away and going behind his buddy, but each time, he came back to me, completing a full circle. It was quite humorous to all of us watching—it was as if his buddy was a barrel he was going around.

*1.22 A & B Luna looks at Jag (A) and when Jag looks back at Luna, they begin to arc toward each other (B).*

Horses clearly skate the edges of each other's bubbles (figs. 1.22 A & B). In a herd, acute spatial awareness is developed so no horse will trip up another as they move as a group. Much of their language addresses these themes of space, as we will see in the rest of the book.

## Contact versus Space

Because they are naturally motivated to preserve safety in the herd, horses value space *more* than they value contact (fig. 1.23). Humans, however, value contact more than we value space. We have hands and a need to *touch*. Our need to touch our horses can supersede our common sense and even get us into trouble. Our love of touch is, in fact, the root of awkwardness and miscommunication with our horses. Horses need to feel their space is respected. A significant amount of their facial communication with us is about their personal space and how they perceive it is being violated. (We'll talk more about these facial expressions in Step 2—p. 41.) But we often just move into their Bubbles of Personal Space and put halters on without asking permission or kiss their muzzles without warning—which is not to say we shouldn't halter, caress, hug, and snuggle with our horses. But in order to be considerate, we might think about doing these things more tactfully. To see when we are invading a horse's bubble, we need to watch the eyes and lips more closely than the ears. Some domestic horses believe there is little point in asking humans to move out of their personal space. They may have been punished in the past if they did. A domestic horse that is used to having his space invaded will only show a hard eye and tight lips to convey: "I don't like this." This horse may stand perfectly still on the cross-ties while you groom and tack him up, but the look on his face is like the look on ours when we are at the doctor getting an injection.

Horses communicate about personal space all the time with their faces. Even when there are only two horses in a herd, sharing a grazing area or piles of hay, their facial gestures will say things like: "I don't mind you close to me," or "You—I don't like you right now, and I don't want you anywhere near me." Horses never want to stop eating, so they first try to negotiate space just by looking at each other (fig. 1.24).

Horses are always asking questions of each other and receiving answers. When one horse approaches another, their Bubbles of Personal Space touch before their bodies have physical contact. *Sharing space* means they are in *overlapping* Bubbles. Personal space is not the entire picture when it comes to Horse Speak, but it is one of the most observable interactions between horses we can witness.

Two spatial relationship Conversations that are easy to identify in horses are *Approach and Retreat* and *Go Away and Come Back*. Although these may sound similar at first, Approach and Retreat is the way a horse—or you—enter another's Bubble. Go Away and Come Back is how a horse asks for some space or Beckons you to come closer.

*1.23 Rocky (on the right) gently tells Jag to move ahead. She reacts to preserve her space, arcing to the side.*

*1.24 In this picture, Rocky asks Vati to move her head away from the hay.*

## Approach and Retreat

Horses are quite specific in their communication with each other, so they do not like people to be vague with them. Approach and Retreat permits the horse or person requesting permission to enter another's space to learn a lot about the other horse or person. Using this template in a Horse Speak Conversation shows a horse that you are trying to be aware of his Bubble of Personal Space. He needs to be allowed a chance to get to know how your Outer Zero (physical) *looks* and how your Inner Zero (emotional) *feels*. Once he sees you watching for a signal to take the pressure off his Bubble of Personal Space, he will trust you more.

Once this Approach and Retreat is ingrained in your behavior, you'll see how basic it is and use it in many interactions and in a general way, as horses do with each other. Approach does not just mean *moving toward* the horse—although this is one expression in the Conversation. Approach is any sort of *asking*. Retreat is the *reply* from the asker once the horse has answered the Approach. Retreat is the release of action, intensity, or asking. It is going back to Zero. Approach is easy for humans to learn as predators, but learning and integrating what I call *Retreat to Zero* takes more effort, which is why learning to *Pause* is so important (figs. 1.25 A & B and see p. 23).

Horses Approach others with tact. They may simply look at, breathe toward, or lean toward each other. Extending the nose in greeting is also an Approach. Retreat can be a complete withdrawal from another's Bubble. Sometimes Retreat may be as subtle as a horse simply pausing for a moment while on Approach. Retreat is any expression of sensitivity to the other horse's Bubble of Personal Space.

*1.25 A & B* *"May I approach you?" I ask. Vati will not look at me, so she is saying, "No" (A). "I will retreat," I say. Vati looks toward me to say, "Thank you" (B).*

## Go Away and Come Back

Horses also negotiate personal space with the *Go Away and Come Back* Conversation, using *Sending* (Go Away) and *Beckoning* (Come Back) messages. We're all familiar with how a horse sends another away. Examples of the Sending language are: pinned ears; grimaced face; raised poll, face aimed toward a *Button* on the other's face, neck, girth, or rump (see p. 36 for discussion of the Horse Speak Buttons); stomping, kicking, striking, and biting (figs. 1.26 A & B). Beckoning language is more subtle: relaxed ears, a steady soft gaze, a lowering of the head or head nods, and soft breathing communication as I described on p. 25, which makes him be soft and welcoming (figs. 1.27 A & B).

*1.26 A & B Dakota's face is giving a Sending message: flat ears, hard eyes, and tight muzzle (A). Rocky is Sending Vati away (B).*

*1.27 A & B Zeke's face is eager Beckoning (A): he wants me to come toward him (A). Dakota's face is soft Beckoning, also inviting me to approach (B).*

As we are starting to understand, horses *Beckon* other horses with facial expressions. The horse that wants company will simply gaze, with a lowered head, at the horse he is welcoming. He may nod his head slightly and use a soft breath message (p. 25). Watch for this expression, which I call *Aw-Shucks*, when you are approaching him: He'll put his muzzle to the ground, asking you to take the pressure off. You

**1.28** *"Aw-Shucks, I just want to be near you."*

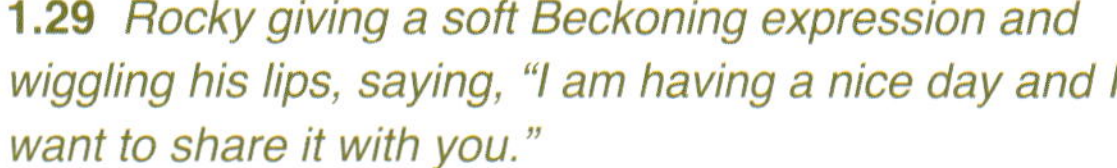

**1.29** *Rocky giving a soft Beckoning expression and wiggling his lips, saying, "I am having a nice day and I want to share it with you."*

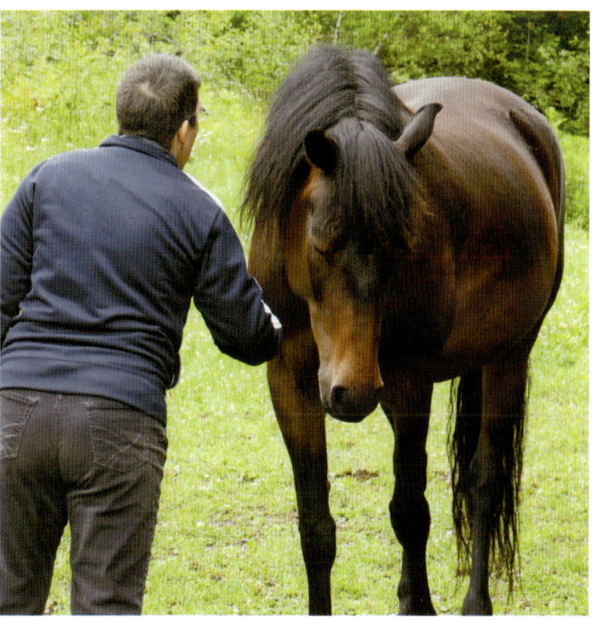

**1.30** *Deferring Space by moving the head off to the side is a friendly and respectful gesture.*

might see his nostrils moving. This bears repeating: Anytime your horse puts his muzzle to the ground and is not eating, whether in a halter, bridle, or at liberty, it is his version of Aw-Shucks (fig. 1.28). Pause in your approach and take a few seconds to do the human version, scuffing your foot and looking down. (Your horse may become so interested in seeing if you will mirror this consistently, he will initiate this Conversation again and again.) Pay attention if your horse looks at you with a soft gaze; if his lips are relaxed or you see them wiggle, which means, "I am having a nice day—join me!" Notice whether his nostrils are "huffing" as if he is trying to "breathe you in" (fig. 1.29).

Another gesture of friendship and welcome is what I call *Deferring Space*. It feels like when someone sitting on a couch moves over to make room as you to sit down beside her. When you walk up to your horse, notice if he gracefully turns his head and neck away from you without moving his legs or any of the rest of his body (fig. 1.30). Also, notice if he is not looking at anything in particular, just "making space" near his shoulder. This supreme gesture of welcome is often misinterpreted by us. We tend to think the horse is turning away in disinterest or is rejecting us in some way, when it is just the opposite!

## THE 13 HORSE SPEAK BUTTONS

To make Horse Speak a little easier to begin to understand as we move through the different Steps in this book, I have reduced the horse's language to 13 "Buttons" located all over his body (fig. 1.31). Think of these Buttons as the ABCs of Horse Speak. Although you will most likely recognize at least a few as "cue zones" that you may have learned about in the past as you acquired horse knowledge and skills, don't get hung up on trying to use them to cue your horse. The hidden reason these cue zones work so well is that these Buttons are embedded in the horse's "code" the moment his momma starts teaching him to communicate with her and with others.

Even though there are actually more than 13 Buttons that horses use when "talking" to each other, the ones I'll teach you for Horse Speak are the easiest to begin to understand, to observe in horses, and to incorporate into your own communication with them. I have broken this subject down to its simplest components in an effort to be clear, but as you read the list below, do consider that horses do not necessarily start with Button number 1 and then proceed systematically through to Button number 13 in order. Instead, they use whatever Buttons they need to convey the message they want to send. In addition, keep in mind that horses can use each Button in both a Beckoning and Sending manner, as well as a simple neutral message that may not require movement as part of the "reply."

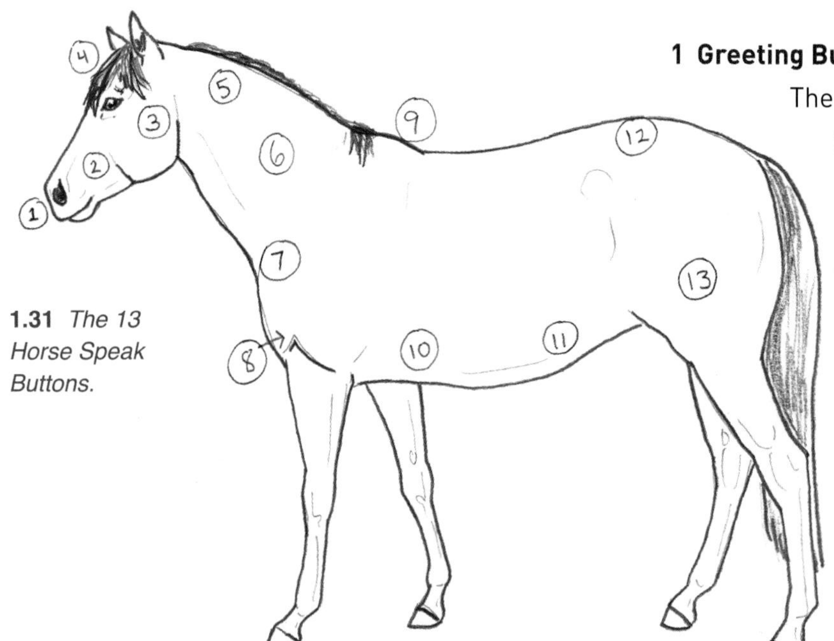

1.31 *The 13 Horse Speak Buttons.*

### 1 Greeting Button (p. 52)

The *Greeting Button* is located on the front of the muzzle and is used in Greeting another (fig. 1.32 A). Similar to a human handshake, much information is exchanged in Greeting via this Button (see p. 52 for more on Greeting). This button is also used for check-ins and as a friendly gesture.

### 2 Play Button (p. 58)

Located on the side of the horse's face just behind and

slightly above where the ring of the bit sits, when touched the *Play Button* simply means, "I want to play with you" (fig. 1.32 B). Pushing on this button (as so many of us have when telling a horse to stop crowding us) is actually a direct invitation to have your space invaded.

### 3  Go Away Face Button (p. 57)

This Button is found at the back of the horse's large, round cheek, under his eye and ear (fig. 1.32 C). It is easy to identify when your horse is wearing a standard leather or nylon halter, as it often aligns with the location of the metal ring just below the halter buckle. Horses use this Button to gain head space from each other. Using it alone (without any other Horse Speak suggestions) simply means, "Give my face a little space." Some horses that are used to being "space invaders" may require several touches at this spot before they understand that you mean *permanently.*

### 4  Friendly Button (p. 122)

Right under the horse's forelock, where the long hairs come to a point, is the *Friendly Button.* The Friendly Button may be easy to recognize if you have a horse that likes to have you rub him there. However, abused or fearful horses will often be very guarded about this spot. I've worked with many damaged horses that "defend" this spot. Horses nuzzle each other on this Button when they have really bonded, and a horse that is bonded to you may rub his head on the front of your body (fig. 1.32 D). The horse is connecting with you using his Friendly Button. Lightly rubbing this spot in reply is a great way to say to the horse you are indeed ready to have a closer friendship with him—but know that you do not have to let him knock you over during this exchange!

### 5  Follow Me Button (p. 120)

The *Follow Me Button* is located on the horse's upper neck about 4 inches down from the poll (fig. 1.32 E). A mother horse will nudge her infant here to tell him it is time to follow her. By pressing your hand into this spot, then removing it completely and beginning to walk away (carefully and softly) you are inviting a horse to follow.

### 6  Mid-Neck Button (p. 64)

The *Mid-Neck Button* is used between horses to demonstrate hierarchy (fig. 1.32 F). It means, "Move your face, neck, and front feet away from my Bubble of Personal Space, *completely.*"

### 7  Shoulder Button (p. 152)

The *Shoulder Button* can also be used to demonstrate hierarchy, but when a horse is aiming solely at this Button rather than the Mid-Neck Button, he is keeping the intensity low and less formal (fig. 1.32 G). He is saying, "Hey, buddy…give me a little room, okay?" This Button is also used in lining up tandem movement, when a pair of horses is *Going Somewhere* (see p. 57) together. In this case, the Button says, "Stick with me, and let's move as one."

1.32

## THE 13 HORSE SPEAK BUTTONS

### 8 Back-Up Button (p. 67)

The *Back-Up Button* is used when one horse wants to claim the space in front of another (fig. 1.32 H). When pressed, it tells a horse he has to back out of the space he is in. The point of the shoulder is where this Button is located.

### 9 Grooming Button (p. 55)

We have all seen horses Groom the top of each other's withers. The *Grooming Button*, located at the withers, is absolutely about connection and affection (fig. 1.32 I).

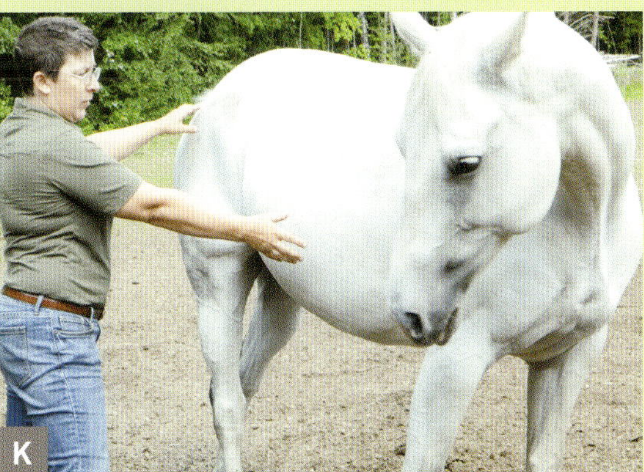

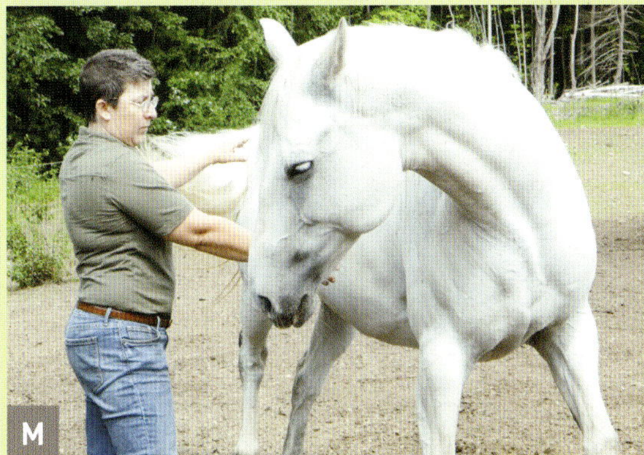

**10 Girth Button (p. 124)**

Horses can ask for space from *front to back* and from *side to side*. The *Girth Button* asks for more space on one the side, and it can also ask for more speed (fig. 1.32 J). For instance, if a horse is Sending another horse away and wants to pick up intensity, he may aim quickly at this Button to say, "Go away, FASTER!" However, this Button is also used as a target area for the young foal's nose. When his nose is at the Girth Button, he is in the center of "Mom," protected by all four hooves, and easily maneuvered. Because of this, the Girth Button can also be used as a connection zone.

**11 Jump-Up Button (p. 125)**

About 6 inches back from the Girth Button is the *Jump-Up Button* (fig. 1.32 K). This vulnerable area is susceptible to predators, and horses may show intense protectiveness here. Aiming at this spot tells a horse to jump up, kick out, or scoot sideways. Stallions spend time wooing a mare in the area of this Button as a way of asking her, "Please, do *not* jump up!"

**12 Hip-Drive Button (p. 131)**

One of the few Buttons that is used almost exclusively as a Sending message, the *Hip-Drive Button* is located at the apex of the pelvic rise at the top of the horse's rump (fig. 1.32 L). Horses have two ways of showing leadership: "Follow me" or "Move in front of me while I drive you." To drive another horse forward is to ultimately tell that horse you are the "Rear Guard" and will deal with predators that are coming up behind.

**13 Yield-Over Button (p. 134)**

The *Yield-Over Button* is found in the hollow above the horse's stifle (fig. 1.32 M). It simply tells the other horse to move his hind end over and out of the way. Aiming at this Button keeps the peace when horses pass by each other and prevents kicking out, essentially saying, "I won't kick if you won't." This Button can also be used to turn the young foal around to face his mom, focusing his attention back on her. When grown horses use it in this way, they are telling each other to stay focused and look to their leader rather than getting worried about what is "out there." To this effect, this Button can help a horse calm down when massaged gently.

# Observing Facial Expressions

Let's now take some time to use our powers of observation to gain an understanding of specific expressions—the beginning of learning to use Horse Speak in your daily interactions at the barn or in the arena.

I watch a horse's expressions carefully because every horse communicates volumes with his face. Humans tend to watch each other's faces. So do horses. We may not even realize our horses say a lot with their faces. The expressions we see depict their internal landscapes: thoughts, feelings, and opinions. As you explore aspects of one horse's face, you will begin to unfold a map to his inner world.

The height of the horse's head, the mouth, nostrils, chin, jaw, eyes, and ears all have stories to tell. The stories get more complex and fascinating when you understand how a horse combines all the facets of his face in order to communicate with other horses and with you. On the pages that follow, I will deconstruct gesture and nuance of the individual parts of the horse's face and note what each means separately. There may be circumstances when a whole message is a combination of expressions and gestures. Over time you will begin to see how they come together to allow a horse to be more descriptive.

Regardless what a horse is feeling, thinking, or communicating, it may only show on his face for a brief moment because a horse returns to a *Neutral* or *Peaceful* expression as soon as he can. Some facial gestures are so fleeting that it took me years to document their meaning.

## Keys to Horse Speak: Step 2

**Head Height (p. 42)**
**Muzzle (p. 42)**
**Nostrils (p. 44)**
**Jaw (p. 45)**
**Eyes (p. 45)**
**Ears (p. 47)**

*2.1  Zeke shows a medium-high head and perky expression, his eager muzzle reaching out.*

*2.2  A lipping, affectionate muzzle.*

## HEAD HEIGHT

As I approach a horse, the first thing I notice is the height of the head. When a horse has a high head, it means he is on alert. This can be due to tension or even just curiosity—either way his adrenaline is up. A low head means, "All's well," and the adrenaline is low. The ultimate goal is for the horse to feel so comfortable that he keeps his head low (fig. 2.1).

## MUZZLE

The muzzle can express great intimacy. Horses Groom their friends (horse or human) with loose lips and a relaxed muzzle. They use their lips to ask other horses if Grooming can continue before they engage their teeth in a harder scratching motion (fig. 2.2). I once met a sweet horse that massaged every person who stood outside his stall with his or her back to him. The gist of it is, if your horse has loose lips and a relaxed muzzle around you, he considers you his friend.

### Lick and Chew

Some communications are easier to notice because the horse holds an expression longer. One such gesture is the *Lick and Chew* (that is, when a horse is *not* eating). I don't believe this motion only means the horse is agreeing with you; in my experience, it really means he is thinking things over. He is digesting what you are trying to say. You can mirror the Lick and Chew by pretending to chew gum. It becomes a Conversation: "You got that?" and "Yes, that is my understanding, as well" (fig. 2.3).

## Biting

A horse can defend or protect his personal space with a bite that clearly says, "Go away." A full bite, however, is the last resort. As we touched on already, horses communicate Go Away in many less harmful ways: ears may go back, the tail may swish, the body may angle a certain way. As the bite gets closer to reality, lips are tight or teeth are bared, eyes get narrow, nostrils pinch, ears flatten back, and a squeal may be heard before the teeth sink in.

When a horse bites a person, it carries the same message as when a horse bites a horse. The horse wants the person to simply get out of his space, Go Away, and leave him alone. It may be the horse believes the person is not listening to his more subtle requests.

Bites from a horse can be painful and dangerous. When people watch and understand a horse's body language more fully, the chance of receiving a full-on bite is greatly lessened.

## Tight Lips

Tight lips say, "I don't like this." If you take a little time to be curious and observe what is happening in that moment, you can create a better Conversation with the horse. Simply noting when lips are tight or not is important. A tight mouth is holding back and holding onto tension. This can also be evident when a horse is concentrating really hard—he may have a tense mouth, similar to a human who is concentrating while taking a test (fig. 2.4).

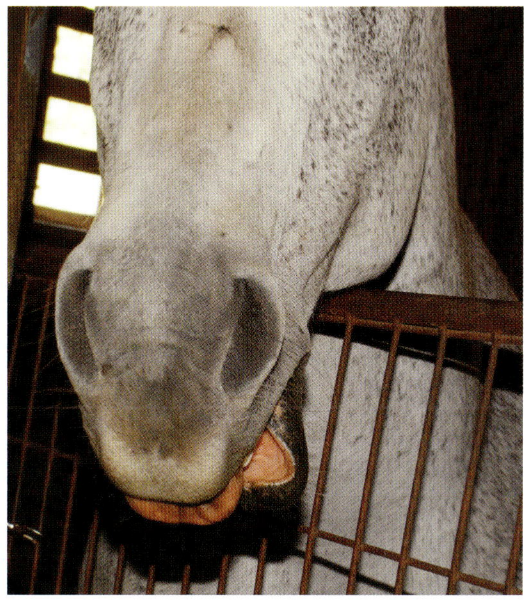

**2.3** The Lick-and-Chew gesture tells us the horse is thinking things over.

**2.4** Tight lips, concentrating ears, and arched poll, show Dakota's intense focus.

*2.5* The curling of the upper lip known as the Flehmen Response can indicate pleasure or pain.

*2.6* A Hard Eye, pinned ears, raised head, and tense nostrils signal displeasure and aggression.

### Curling the Lip

Curling of the upper lip is another gesture easily recognized—we know it as the Flehmen Response. Horses may do this in the presence of strong odors, like coffee or cigarettes. Stallions do it before they breed a mare in order to carry her scent deeply into their nostrils. This same curling of the upper lip can also indicate pain (fig. 2.5). There is a point between the center of the nostrils that is the pressure point for the limbic system. It extends from the center point between the nostrils all the way through to a corresponding point under the upper lip at the gum line. Massaging either location stimulates the release of endorphins, so by curling his own lip when he is in pain, the horse is self-medicating.

## CHIN

A horse that is very tense or fearful will often go beyond tight lips and pucker up his chin, as well. A relaxed chin, on the other hand, is a sure sign of confidence.

## NOSTRILS

The nostrils change shape and size as the horse uses different Breath Messages (see pp. 24–28). Generally, relaxed nostrils mean a calm horse, and hard, small, tense, or puckered nostrils (that look as though the horse smelled something bad) show he is having negative feelings (fig. 2.6). Sometimes you can change the proximity of items in the horse's environment and notice if your efforts cause him to relax his nostrils. You can also gently stroke the nostrils, pulling gently on the edges to soften them.

## JAW

The jaw of the horse is located directly under the eye—you can easily detect its circular shape. Just like us, a horse can clamp his jaw, which means the horse is having tense feelings or bracing or may be holding his breath. Back on p. 26 we discussed how the Yawn can be a Breath Message and tension reliever—it is a way for a horse to release and relax his jaw (fig. 2.7).

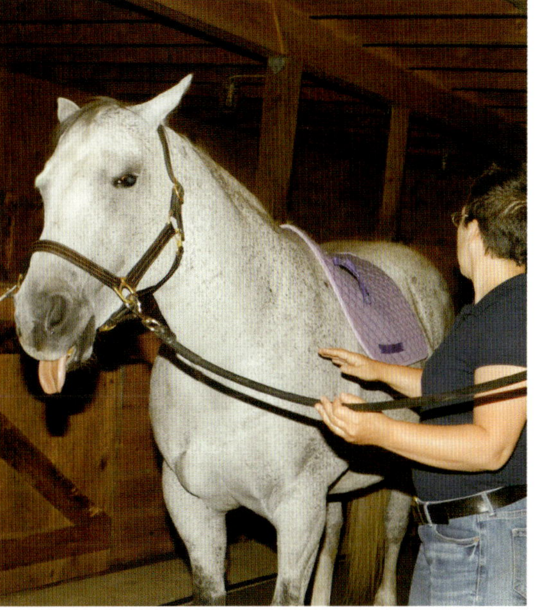

*2.7* This Yawn is a release of tension.

## EYES

Eyes subtly communicate emotion. Here are some of the ways you can read them:

- A calm horse has a *Soft Eye* because the muscles around the eye are relaxed. Watch when a new dog or cat comes into the stable. You can see curiosity in your horse's whole face (mouth soft, ears forward, nostrils flared) but the eye is relaxed (fig. 2.8).

- Wrinkles or puffiness under the eye tells you the horse is in physical or emotional pain. If the upper eyelids are wrinkled or tented the horse is becoming extremely distressed.

- When the eyes are wide and showing whites, the horse is panicked.

*2.8* A low head, perky ears, chewing, open nostrils, and reaching lips mean the horse is asking to come closer.

- Tension around the eyes says, "I don't like this." We often see this when the vet or farrier arrives.

- One horse may give the horse next to him something I call a *Hard Eye,* which says, "Move over." As humans, we joke about giving someone "stink eye," like an old-fashioned gunslinger's look. It is the same idea. A horse can gain some space using just "eye pressure"—it is enough to move another horse. When an eye is not enough, he may add an ear and arch his poll—note that the horse that arches his poll, no matter how subtly, is willing to take the Conversation to the next level.

- I have observed horses turning each other using what I call *Laser Beam Eye.* This when they look each other right in the eye with one intending to change the other's path of travel. I have seen many horses redirect the eye path of another horse, which then redirects where that horse's body goes. You can mirror Laser Beam Eye: Look a horse right in the eyes and imagine a beam of light shooting out of your eyes and blocking his forward momentum.

## LASER BEAM EYE
### IN PRACTICE

Once I was being charged by a severely damaged horse at a clinic I was attending. I was on my own in the round pen, waving my crop, and making my short body as large as I could, but this horse kept coming at me. So I looked him right in the eye and imagined a beam of light cutting across his path in order to change his direction. It worked. The clinic participants wondered how I had done it.

Not long after my first Laser Beam Eye experience, I was consulting at a rescue and there was a band of rowdy geldings in the next paddock. I couldn't make any progress with the horse I was talking to with all this acting-up right next door. So I decided to watch the group for a moment. The leader of the pack was easy to pick out, and I applied my Laser Beam Eye on him, cutting him off from the rest of his group. After I did this, he came up to the fence and looked at me as if to say, "Where on earth did you learn that?" He knew I was now in charge.

## Blinking

You rarely see a horse with his eyes completely closed. Sometimes, when I am rehabilitating a horse, I will nod my head and Blink slowly while looking at his eyes. I am gratified when he Blinks back. I believe Blinking is a sign of thinking (like Lick and Chew—see p. 42), and it tells me he is taking a moment to mull things over. I often use the saying, "Blinking is thinking." Slow Blinking can also show affection. When you experience this sweet gesture, take a moment to slowly Blink back. Blinking at horses helps them to relax around you (fig. 2.9).

## EARS

We tend to watch horses' ears more than other facial features because ears move and attract our attention. The ears indicate where the horse's mind is going; what the horse is paying attention to.

- One ear, or both ears backward toward the rider or driver says, "I'm paying attention to you." One ear back and the other ear forward at the same time says, "I'm also attending to the work at hand" (fig. 2.10).

- If the ears are predominantly forward but one is twitching forward and backward, your horse has *Curious Ears*. These ears are combined with other characteristics of a curious expression, such as the eyes discussed on p. 45. A curious face looks a lot like the face of a horse begging for a treat.

*2.9*  *Blinking, low heads, soft ears, looking right at the camera says, "I feel fine, how are you?"*

*2.10*  *In this photo we see one ear focused ahead, one behind, and a Hard Eye: the horse is chewing it over, thinking, "Do I stay or do I move?"*

*2.11* Airplane ears, low head, and a soft muzzle are welcoming.

*2.12* Inward ears plus Lick and Chew mean the horse is quietly lost in thought.

*2.13* Scouting ears with "tented" eyes that bulge indicate extreme alertness.

*2.14* A horse might "shake out stress" by shaking his ears the way a dog shakes off water.

- A horse that has *Airplane Ears* is relaxed and confident: this flat-sideways ear position also draws or Beckons another horse or human (fig. 2.11).

- There is a gentle state of mind I call *Inward Ears*: ears held softly backward and a little downward (fig. 2.12). These ears communicate that the horse is not really paying attention to much of anything on the "outside." Mares may hold this Inward Ear position while they are nursing. Notice if your horse has Inward Ears when you walk next to him. It indicates a quiet or peace within, usually evident in times of unity or friendship.

- *Scouting* Ears are held straight up but only for moments at a time. The Sentry horse (see p. 60), in charge of looking for danger in a herd, will have Scouting Ears when a threat is perceived. Scouting Ears also indicate the horse is ready to go on the offensive. A stallion challenging another has Scouting Ears and an elevated poll to demonstrate he is a "big, bad horse," compared to his foe. Horses working cattle show Scouting Ears just before the ears are laid back into "attack" mode to move the cattle. You can also often see this ear shift from Scouting to "attack mode" right before a horse takes off over a jump (fig. 2.13).

- Horses "let go" of stress by shaking their ears (as they also do by Yawning—see p. 26). A horse may shake his ears like a dog shakes off water. When he does this, he is literally shaking off a disagreement or a disagreeable event (fig. 2.14).

- I've seen horses play and be silly all by themselves, and their ears "talk" while they do this. Watch two horses eating hay out of each other's mouths, for example. Their ears will twitch and twirl. I believe this is the equine equivalent of laughing (fig. 2.15).

*2.15 When horses eat hay out of each other's mouths, their ears often twitch and twirl—the equine equivalent of laughing.*

2.16 *Here we see relaxed, neutral ears.*

■ No matter what emotion you see in the face of your horse, he will always return to a state of calm and peacefulness as soon as possible: *Neutral Ears* (fig. 2.16).

## LEARNING TO READ

Now that you have the tools you need to read a horse's facial expressions, including the height of the head, muzzle, chin, nostrils, jaw, eyes, and ears, you are ready to take the next step in learning to use Horse Speak (figs. 2.17 A & B). You can observe and read, and now we'll examine how to respond in a way the horse will understand. This is the beginning of Conversation.

2.17 A & B *The mare's clamped jaw, concentrating ears, Hard Eye, and tight nostril say, "I am stressed and not sure what to do" (A). When her jaw unclamps with ears sideways and eyes more present, then it is, "I know what to do now, I am okay" (B).*

# The Four Gs
## of Horse Speak and the
# Greeting Ritual

The Four Gs of Horse Speak are: *Greeting*, *Going Somewhere*, *Grooming*, and *Gone*.

- Greeting is how one horse meets another horse or human, or other animal for that matter (see p. 52).

- Going Somewhere (see p. 57) entails movement: a horse or human moving another, a horse or human being moved by another, or a horse moving together with another. This movement can be minor (yielding the head from your Bubble of Personal Space) or major (changing locations in the arena or pasture).

- Grooming (see p. 91) is a mutual invitation for touch. A willingness to Groom or be Groomed on the withers or crest or other places (with the lips or teeth between horses and with our hands and fingers when we're involved) indicates an intimate connection.

- Gone (see p. 129) is the horse's way of saying, "I'm finished with this," or flat out, "No!" It is a period at the end of a sentence; a break in Conversation. It is an important element that you both need to be able to recognize and to use when talking to your horse in Horse Speak.

In this chapter, we'll learn about the *first* G: Greeting.

**Keys** to
Horse Speak:
Step 3
**Processes of Three (p. 52)**
**Knuckle Touches (p. 52)**
**Copycat (p. 52)**
**Greeting Ritual (p. 53)**
**Rock the Baby (p. 55)**

## GREETING

Humans greet each other with a formal handshake the first time they meet, and horses have a similar system. The Greeting Ritual is the basic platform I have created to teach humans how Conversations with horses can exist.

The Greeting Ritual consists of *three separate moments* in which horses that are meeting touch noses on the Greeting Button (see p. 36). The speed at which they may perform these three touches varies from lightning-fast to very slow. The reason for three official touches is simple: there is much to say in a first, formal greeting, and it takes two subsequent touches to sort it all out. Plus, I find that horses learn about the world around them in processes of three or more.

### First Touch: Formal Greeting, "Hello," and Copycat

A horse's pecking order is different from a dog's, for example, because horses are concerned with how to run and move together in case of emergency. In a horse's world, any moment could bring danger, and the more alpha a horse is in pecking order, the more responsible that horse is for fending off attackers. In dog psychology, the alpha dog calls the shots, but in horse psychology, to lead is to be responsible for the welfare of those that are weaker. The only difference in horse social order occurs between stallions. A herd is typically made up of grandmothers, aunts, mothers, and daughters. There is one dominant stallion, and he not only guards his ladies from other rogue stallions, but from mountain lions, wolves, and bears. Stallions that have no mares band together in bachelor herds, and although they can enjoy some rowdy play, they tend to develop strong emotional bonds with each other and follow the same herd dynamics as any other herd.

Domestic horses are bought and sold and moved from barn to barn, and have to deal with new herd members often. Sometimes they are not turned out directly with other horses and can only socialize over a fence, if at all. But the formal Greeting Ritual remains essential for horses who are just meeting each other to be able to sort out who will tell who what to do.

The First Touch is much like a formal human handshake (fig. 3.1). It is the "Hello" followed by an immediate question: "Where

*3.1 The First Touch in the Greeting Ritual: "Hello!" and "What's your status?"*

are you in the pecking order?" Horses size each other up and assess very quickly as much as they can about each other's herd status. There is much more to it than just calling one "Alpha," because a healthy herd has many diverse roles that get played out: the "Peacemaker," the "Bully," the "Sentry," and the "Joker," to name a few.

Immediately after the first nose touch, one horse will make a move in one direction or the other, which can be as subtle as a shift of the head position. The question is: "If I want to go that way, will you follow?" I call this *Copycat*, because in a friendly exchange, one horse will Copycat the other's movement to indicate, "Yes, I will follow you."

## Second Touch: Getting to Know You and Copycat

After the delicate first encounter, there will be a second nose touch, usually with a deep inhale. This touch is a getting-to-know-you breath and will usually be accompanied by another Copycat gesture to confirm the leader/follower roles. This step can be performed slowly or very quickly depending on the personalities of the horses.

## Third Touch: What's Next?

Now the third and final nose touch can occur. This touch is the lead to other avenues of contact or the opportunity for the horses to break contact and go their separate ways. The Third Touch is important and in a superfast greeting ritual—such as may be likely to occur between two highly charged horses—this final straw can erupt into squealing, striking, or other displays of contest or play. However, in relaxed horses this touch leads to the other Gs: Going Somewhere, Grooming, or even Gone. This is when the peaceful herd dynamic that horses desire most starts to emerge.

Luckily for us, horses seem to be impressed when humans figure out to extend their knuckles in *one* "Hello" touch, and they are happy to welcome us into their world even without the entire formal Greeting Ritual. However, when a person can perform the whole Greeting Ritual, there is so much information exchanged that both parties will feel a much more intimate and intense understanding of each other.

### *Conversation:* Greeting Ritual

**1** Knuckle Touch to the horse's Greeting Button to say, "Hello," followed by an obvious turn to the side to see if the horse will Copycat your movement and offer to be your follower (figs. 3.2 A & B). The Knuckle Touch should be made with your hand in a soft fist, knuckles up.

*3.2 A–F* First Knuckle Touch: A cautious, "Hello," staying mindful of our Bubbles of Personal Space (A). Copycat: I offer to lead; Vati cocks an ear in that direction (B). Second Knuckle Touch: A more comfortable touch as we are "getting to know" each other. Vati offers to sniff my hand and begins to cock her ear to the side (C). Copycat again: Vati keeps sniffing my hand as I look again to the side, indicating she is trusting me more (D). Third Knuckle Touch: We breathe softly at each other… (E) …and then ask, "What's next?" (F).

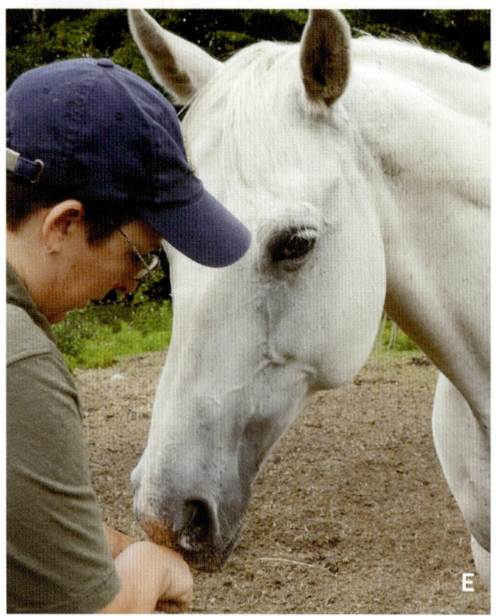

**❷** A second Knuckle Touch to say, "Getting to know you!" followed by one more turn to the side to confirm the horse will offer to follow you (figs. 3.2 C & D).

**❸** A third Knuckle Touch to say, "What's next?" with soft breathing. This could lead to you going somewhere together, grooming, or separating peacefully (figs. 3.2 E & F). The third touch is where the next level of Conversation begins. This is where some inherent differences in what humans and horses value can lead to misinterpretations of actions (figs. 3.3 A & B). Horses value space more than touch, but we value touch more than space (at least with animals).

## ROCKING TOGETHER

No doubt, you've seen horse friends express connection with each other by grooming each other with a rhythmic toothy scratching of each other's withers and back. We are going to discuss Grooming at greater length later in the book (see p. 91) but for now you should know that a horse may invite Grooming from a human by lingering at the *third* Knuckle Touch in the Greeting Ritual. When this happens, naturally reach to scratch the withers area. I recommend placing your other hand, knuckles up, below the horse's muzzle so he has something to target if he feels like lipping you in return.

When you watch two horses scratching each other's withers with their teeth, you will see them rocking together slightly. There is a sweet Conversation you can mimic—I call it *Rock the Baby.* It takes advantage of this friendly area just where the withers slope toward the horse's back—the Grooming Button (see p. 39). If you are comfortable in your relationship, you can Rock the Baby on the horse's withers while he is at liberty in his stall or pasture. Later, we will discuss doing it with a halter and lead rope (see p. 79).

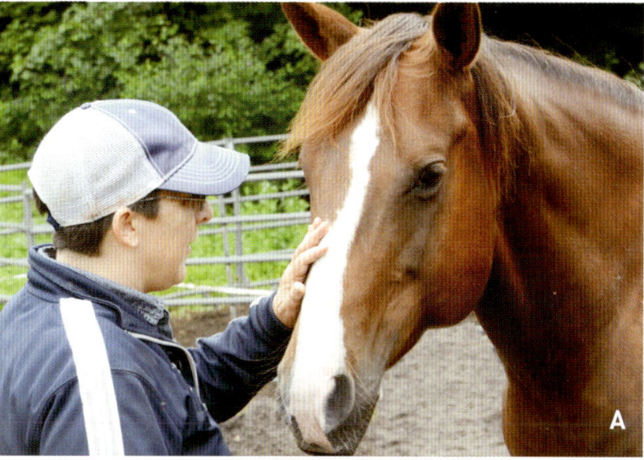

*3.3 A & B*
*Some horses naturally enjoy friendly touch (A). Others are more reserved about touch—horses value space over touch (B).*

*Conversation:* **Rock the Baby**

**❶** Stand with Inner Zero (see p. 8) at the horse's girth line, facing the same direction your horse is looking.

**❷** Place your hand closest to the horse across his withers.

**❸** Now slowly shift your weight from one foot to the other. Start connecting your shift to your in-breath and out-breath.

**❹** With your hand, feel the effect you are having on the horse's body. You'll begin to feel his weight shift from one front leg to the other. You can begin to quietly encourage his internal "rock": Slightly increase your movement, perhaps by swinging your hips slightly with your weight shift. The idea is not to forcibly rock him but to let him *join in the movement* if he wishes.

The Rock the Baby Conversation connects your horse's sense of balance to your own. Anytime you foster balance with your horse you create mutual balance. When you create physical well-being (via balance) with a horse, you are also creating emotional well-being.

I must note that many horses will choose Going Somewhere rather than Grooming directly following the Greeting Ritual, so we are going to explore that "G" in more detail first (see p. 57). However, we as humans who crave touch usually want to put our hands on our horses as soon as possible. If we can delay our need for physical comfort and follow the horses' protocol, we will create a deep sense of trust in our horses. Even a horse who seems to *want* touch, who is either conditioned to be used to it or comes across like a sort of giant puppy, rubbing his head all over you, will benefit from your restraint and from following the Going Somewhere part of the ritual first. The reason is simple: Establishing who leads whom, and who follows whom is more essential to our mutual trust and safety and will generate more authentic affection from the horse than touch. He will enjoy your clarity of purpose.

Luckily, the Going Somewhere part of the Four Gs of Horse Speak can be enacted simply, painlessly, and with great effectiveness—even across a fence or over a stall door.

## GOING SOMEWHERE PART ONE

### Go Away Face Button and Play Button

There is a Button toward the back of the horse's cheek, which I call the *Go Away Face Button* (see p. 37). Horses use this Button by either indicating toward it in the air or by directly nudging it. This is one of the first Buttons "Momma" would have taught her newborn foal. As we discussed in Step 1, In Horse Speak, this Button simply means, "Move your face away from my Bubble of Personal Space." Since horses have such long necks, it is not always necessary for a horse to move completely away from the spot he is standing in order to give another horse space. By simply yielding his face to the side, he can give adequate room for another horse to feel respected.

You can see this "bubble dance" around horses' faces at the water tank, over a pile of hay being shared, or in manmade situations, such as when horses are in harness together or tied near each other. In these scenarios it may not be feasible or possible for horses to actually move away from each other completely, so simply yielding head space is enough to keep the peace.

## Keys to Horse Speak: Step 4

**Go Away Face Button (p. 57)**
**Play Button (p. 58)**
**Scanning the Horizon (p. 59)**
**Sentry (p. 60)**
**The Forehand (p. 62)**
**Don't Pick Up/Pick Up Feet (p. 63)**
**The Mid-Neck Button (p. 64)**
**Move Your Feet Over (p. 65)**
**"O" Posture (p. 66)**
**"X" Posture (p. 68)**
**Core Energy (p. 68)**
**Blocking Forward Movement (p. 70)**

**4.1 A & B**
*Often it is not necessary to physically push the Go Away Face Button (A), simply pointing at the Button is enough (B).*

**4.2** *Rocky curls his lip playfully as I touch the Play Button. People often push here when trying to get a horse out of their space—which sends mixed messages to the horse.*

For our purposes, knowing about this Button is key to gaining a horse's respect for our own Bubble of Personal Space, without necessarily requiring him to use his feet and step away. Since this is a Button that *all* horses use, and its use is in context with a calmer space-claiming request, including it in Horse Speak taps into the horse's innate sense of calm Conversation.

The first step to asking a horse to engage in Going Somewhere is to simply use the Go Away Face Button and request that he move his face to the side, giving your Bubble more space (figs. 4.1 A & B).

Interestingly, horses that tend to be "space invaders" often have their muzzles or noses pushed away by their handlers (sometimes with a lot of force). However, the area between the mouth and the cheek bone, right above where the bit ring sits, is the *Play Button* (p. 36)! Horses will nudge and nip at each other's noses and muzzles to indicate they want to play, and so touching there indicates it is perfectly fine for them to invade your Bubble of Personal Space (fig. 4.2).

This major misunderstanding between our two species causes much frustration and even head-shyness in our horses. By pushing on the Play Button but intending to send the horse's face away, we end up frustrated with a horse that is doing exactly what we said he *could* do—invade our bubble and play games with us.

Aim your hand higher up on his cheek, directly under his eye, and use a firm finger when you push the Go Away Face Button. This will get excellent results because the horse will understand what this means: move his face over and keep it away from your bubble. When your horse is used to putting his face in your space, you will have to reinforce this Button several times before he understands that you mean for him to give you space *permanently*. In contrast, if you push on the side of his mouth or nose, you will see that he swings his face into your bubble very soon after, ready to play.

 In some cases, a horse may be very defensive and pin his ears when you ask him to move his face away. If this happens, keep your distance for your own safety, and use a crop with a soft end or a light item that swings, such as some baling twine, toward the Go Away Face Button from a reasonable distance. *As soon* as your horse moves his head *a little bit away*, go to Zero and leave. If a horse is this defensive, there are reasons. You do not want to enter into a pushing contest with a defensive horse, and you do not need to come at this phase with an improper attitude of needing to win, prove you're boss, or any other aggressive feelings. He is already defensive, so it won't help to make him *more* defensive. Remind your-self that you are emulating his mother's first nudge, and that this Button is so engrained in Horse Speak that it has its own powerful meaning to him.

I have had incredible success with even violently defensive horses by staying at Zero and repeating my request at regular intervals after the "Hello" Knuckle Touch. As long as you thank him for complying in any way, let it go and walk away once he has made even the smallest try, this Button works miracles. The reason it can be so transformative is simple: you are not asking for any other parts of his body to yield space, you are quitting fast and returning to Zero, which imitates how horses make requests. You are also using a very intimate Button in Horse Speak. Remem-ber, this is the Button horses use for low-impact, calm Conversation *in most cases*. This Button builds both trust and respect with very little effort on either side of the fence, and it will convince the horse that you are indeed trying to talk to him in Horse Speak.

## Scanning the Horizon and Sentry

Before you are ready to move the horse's front feet over and begin the ritual of Going Somewhere with the forehand, you need to advance the Copycat game we learned about in the Greeting Ritual (see p. 53) to the next level. Horses use a gesture I call *Scanning the Horizon*. This simply means that the Alpha horses will look far and wide on occasion and decide if there are threats or not (fig. 4.3). At regular intervals in the day, or anytime there is a disturbance, herd leaders will lift their heads high, stop chewing, perk their ears up, and sniff deeply. When one horse wants to encourage another to follow him or wants to reassure another horse that he is safe and all is well, the leader will Scan the Horizon *for no real reason* other than to prove to the "weaker" member of the herd that he is looking out for everyone, and should be listened to. If all is clear, the horse simply lowers his head and returns to eating, often with a nose-clearing snort, which I'll discuss next.

The advanced stage of Scanning the Horizon includes the act of *Sentry.* Usually one horse plays Sentry in a herd. This is the singular animal that makes the decision to run or not. The Sentry will often be more dramatic in his assessment of the environment and include a loud blow through the nose in the direction of any disturbance—the *Sentry Breath* (see p. 27). The loud blow happens for a few reasons: First, this clears the nose of debris so the horse can get a better sniff of whatever is out there; second, the sound is startling to hear and causes the other horses to look up and pay attention while warning the potential threat that the Sentry is on to them.

We've already talked about the Sentry Breath in this book—"blowing away the bogeyman," as I call it. Making a loud blowing sound in the direction a horse is spooking at (and you may never know what it was the horse saw or smelled, so just assume that horse-eating squirrels are loose again!) causes the horse to believe you are alert to the environment and watching out for his well-being. (I have actu-

*4.3 Rocky Scans the Horizon to show Vati he is looking out for her.*

ally had horses at liberty come hide behind me after I blew away their bogeymen.) After a horse blows the Sentry Breath, if all is clear, he will visibly relax by lowering his head, Licking and Chewing (see p. 42)—this tells the herd that everything's fine. When you blow a Sentry Breath, you will need to return to Zero, too, and even pretend to chew gum to tell your horse that all is clear.

Scanning the Horizon and Sentry Breath are important to know about before you enter into the Going Somewhere ritual with a horse because staying safe is what horses value and are concerned about. A great deal of what horses talk about involves their safety, and the herd leaders spend a significant amount of time reas-suring the other members of their ability to preserve peace in the band.

## *Conversation:* Scanning the Horizon and Sentry

**❶** Approach your horse, and use a Knuckle Touch on the Greeting Button to check-in or say, "Hello."

**❷** Visibly Scan the Horizon in one direction or another, and use a Sentry Breath when you do (fig. 4.4 A). If you have been regularly playing Copycat with your horse, he may look in the direction that you look in. Remember, horses are very subtle; he may only lean or cock an ear in response.

**❸** When you blow a Sentry Breath, your horse may really perk up his ears and stare hard at your "bogeyman" (fig. 4.4 B).

**❹** Dramatically look back at your horse, "chew gum," and even say out loud, "All's clear!" taking a moment to watch his response (fig. 4.4 C). It is not uncommon for horses to really release some tension in this moment, drooping their lips, letting their ears hang to the sides, lowering their heads, and Yawning (see p. 26).

Do not be surprised, however, if he suddenly looks off at bogeymen everywhere! Your horse is really going to love the fact that you just told him you will Scan the Horizon and take care of any perceived threat. You have just convinced him you are up to the challenge of being his leader, which in his mind means you will protect him from the ills of this world. If you happen to be dealing with a very stoic horse who typically comes across as everyone's babysitter, you are offering him a well-deserved break from the stress of being the caretaker of all. If you are dealing with a very timid horse, he will show visible relief—he may even seem overwhelmed by it—once you have told him in Horse Speak that you will watch his back.

*4.4 A–B  I Scan the Horizon to show Vati I will look out for her, and she cocks an ear in the direction I am looking (A). And I perform the human version of Sentry Breath as I stand beside Jag (B). Finally, "All's clear," lowering my head, Licking and Chewing (C).*

# GOING SOMEWHERE PART TWO

## The Forehand

Respect, in the horse's world, is regard for one another's Bubble of Personal Space. Horses negotiate their comfort zones not just by asking other horses to leave, but also by inviting others to come closer and Share Space. Earlier I explored some of the Sending messages horses use to define or defend their space, as well as Beckoning messages expressed with the face (see p. 33). Like the face, the forehand also has a concentration of expressions that negotiate personal space. There are Buttons on the neck, shoulder, and legs, which also effectively communicate Sending and Beckoning messages. Since the forehand is close to the front of a horse's bubble, messages can be emphatic and effective.

### Thresholds

For a horse, every single *threshold* (the point or level at which something begins or changes) is potentially dangerous. In the wild, the edges of things, such as treelines, riverbeds, boulders, and crossroads could be hiding predators waiting to pounce. For the horse, there is also not much difference between a natural threshold and a manmade one, such as a door or gate or the edge of a lane. Years of evolution have given him the reflexes to make a dash for it should a shadow in a corner turn out to be a lion. Remember always: in the wild it *could* be a lion, and those reflexes don't get shut off just because we find them inconvenient.

When you enter into the ritual of Going Somewhere, keep in mind that in your horse's world this means you are going to learn to travel in harmony together and that you are offering to watch out for the shadows and thresholds, constantly giving him feedback that all is well, the way another horse would.

### Defining Space with the Front Legs

Horses use their front legs to help define their space. A leg *strike*, for example, is an emphatic gesture that another horse is in their space and had better back off. *Stomping* is the exclamation point to the request for space. Stomping also communicates disapproval. Some horses strike or stomp on the cross-ties or when asked to stand under saddle.

Have you ever seen a horse scratching his head or mouth on his leg and kneecap? An equine acupuncturist taught me that because main points of the Lung Meridian are along the leg, endorphins are released when the kneecaps are rubbed or scratched (fig. 4.5). You might see foals nibbling at their mothers' kneecaps or each other's kneecaps. When your horse "dives" toward the ground to scratch his face on his leg and kneecap, there is more going on than him just having an itchy face. He is taking a moment to make himself feel better. Similarly, you can massage your horse's kneecaps to help him unwind after a performance.

One horse can make another horse pick up a front hoof by nibbling on the other's front-leg chestnut. However, making another horse pick up the front feet can mean one of two completely different things: In one instance, the horse nibbling on the chestnut is trying to knock the other horse off balance. You can think of this as a game of "King of the Hill." I've seen young geldings playing this, taking turns—almost bowing to each other—as one causes the front leg of the other one to be picked up.

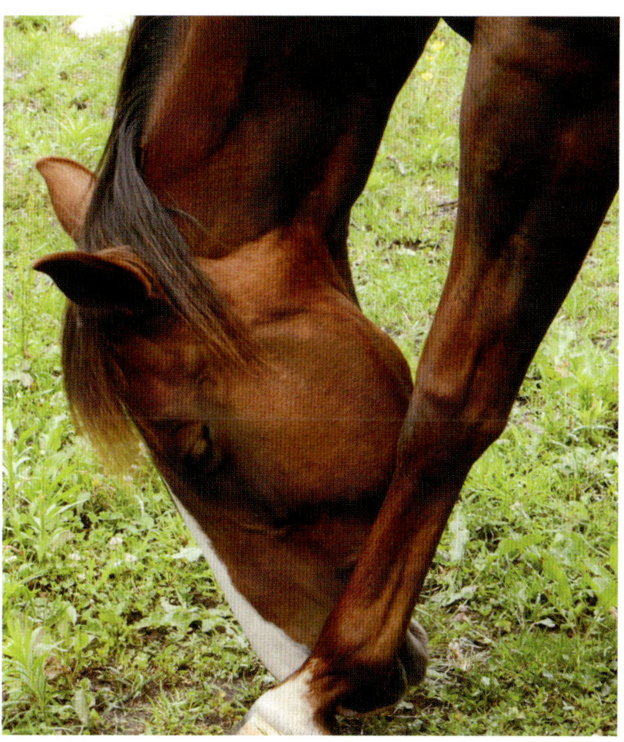

4.5 Jag scratches her nose up and down her front leg, releasing endorphins as she does so.

The other Conversation that happens with picking up the front feet is when a stallion nibbles on the front legs and chestnuts of a mare he wants to breed. Stallions often combine this with nibbling up and down the front legs, as well as licking the mare's legs. The objective for the stallion is to make sure the mare will keep all feet on the ground and not kick him when he mounts her.

So there are two distinctly different objectives with the exact same gesture: One, to actually get the other horse to *pick up a foot* in order to throw his balance off, and the other to encourage a horse to *not* pick up the foot but instead stay balanced, with all four feet on the ground.

When you pick up a horse's foot, he can draw two completely different conclusions based on these almost opposite objectives. When you have a horse that doesn't like to pick up his front feet, he may be thinking that you're playing King of the Hill with him. The horse that picks up his feet for you may think he is losing the "game." This in turn might cause him to not be submissive because he really wants to win by *not* picking up his foot. If he strikes or paws when you are trying to clean his feet, this is a message of hierarchy: he considers himself higher in the herd order; therefore, he doesn't agree to losing the game and losing herd status.

Reverse psychology works well with these horses. Take time to have the Conversation a stallion would with the mare to sort it out.

**❶** Place both hands on either side of one of your horse's front legs, as if you were holding a sandwich.

**❷** Stroke down the leg *without* asking him to pick the hoof up. Use long strokes, from the shoulder on down. This indicates you really *don't* want him to pick up his foot: a Conversation the horse understands. If he *does* lift the foot, don't clean it out with the hoofpick. Stay with the Conversation you've started.

**❸** Eventually, *do* ask for his hoof to lift with a touch on the chesnut. Circle the foot gently in an up-and-around motion. Massage the ankle and fetlock. Scratch the bulbs of the heel. Put the foot back down, and go back to stroking the leg as in Step 2. In no time, you will change the Conversation from one of hierarchy to one of cooperation.

The next part of Going Somewhere includes the horse's front feet. It is not necessary to rush into this phase, however. If you are having a good Conversation about moving his face over, and he is lowering his head, Yawning, relaxing, or holding his face very, very still in the "away" position, he is clearly not only understanding that you are you talking to him using Horse Speak, but he is adjusting his ideas about you and softening up inside himself as a result.

## Move Your Feet Over and Mid-Neck Button

What you now want to include with the Go Away Face Button is the Mid-Neck Button, which is located on the lower part of the muscled part of the neck (and if he resists this, you can try the Shoulder Button—see p. 38). Both these locations are Buttons that mean,

*4.6 A–C Rocky points at the Mid-Neck Button to move Vati's front end over (A). Now they are going somewhere together (B). Jag sends Vati away, aiming at the Mid-Neck Button (C).*

"Move your front feet over." These are methods of *Sending* (see p. 33). How to choose between them? Amongst horses, the most available Button in relation to positon is used (figs. 4.6 A–C).

## *Conversation:* Move Your Feet Over

❶ Some horses move their front feet over better from pointing your pointy finger into the neck, and others are more willing from the point of the shoulder. Whichever works best is fine, but you must include the Go Away Face Button on the cheek that you have already practiced (figs. 4.7 A & B). Stand at your horse's neck, facing him, as you ask him to move his front end over.

At this juncture, you should be able to stay at Zero, hold an attitude of curiosity, and be willing to talk to your horse about how he would feel about politely moving his front feet to the side *along with* yielding his face. Between horses, this is a more demonstrative sort of "Bubble respect," and can even create a squabble over the hay pile. It is not uncommon for a horse to be concerned that you are going to squabble with him and become pushy. That's why it is important to only ask for a tiny step, return to zero, and even blow another Sentry Breath or simply breathe in a deep manner to show him that you are only asking for a new Bubble Conversation, not a confrontation.

When you can ask for him to yield to the Go Away Face Button and move his front feet over, too—and he is beginning to realize that you are, in fact, still just talking to him, not trying to force something or teach something (that he already knows)—he should start to relax his face, drop his head, and breathe more deeply. These things say that he is understanding you.

Try not to drill, but do repeat the Conversation in intervals until you feel that you are asking from a place of Zero, with curiosity and care, and he is responding from an interested and more respectful point of view. What you are telling him is the same thing his mother told him in the birthing stall: "You and I have to move around

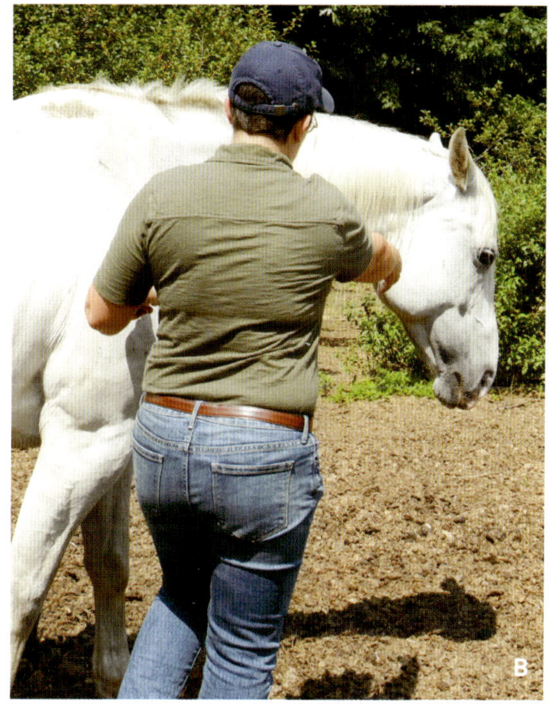

**4.7 A & B**
*Here I am pointing at the Go Away Face and Mid-Neck Buttons (A). Other horses respond better when you point at the Go Away Face and Shoulder Buttons as I am doing with Vati (B).*

each other in this little space. I will set it up so that we are both in good 'Bubbles' so no one gets stepped on. Follow my lead, and all will be well."

## Beckoning and "O" Posture

Next, you can try stepping backward, and inviting his face and front feet to step toward you. Again, you are still influencing the horse's face and neck and forehand, but now you are practicing *Beckoning* (see p. 33). You do this by rounding your shoulders, bringing your hands together in front of your belly, and holding your hands in a welcoming position. Your overall body shape will be in an "O" (figs. 4.8 A–C).

Taking the shape of an "O" is welcoming, and we universally and unconsciously perform this body language when patting our thighs to call a dog or reach down

to welcome a toddler onto our laps. By using a Beckoning "O" posture, we tell the Greeting Button on the horse's muzzle to *come to us*. Again, you are only influencing the *front end* of the horse—this is important to remember at this time because the horse's front end is used most often during any conflicts amongst the herd. Horses drive the forehand away in big conflicts, sometimes causing the weaker member to perform a rollback to move as fast as possible away from the aggressor. (In a conflict, the neck and shoulder Buttons are the most targeted, and if the conflict is between two defensive stallions, then the scuffle involves more Buttons in battle.)

It is natural for horses to be concerned that giving us access to these Buttons means we could be aggressive with

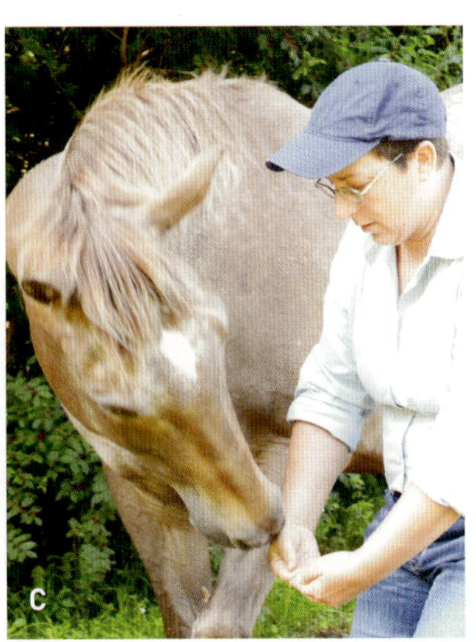

*4.8 A–C I use "O" Posture with Zeke as we share space at Zero (A), and my "O" Posture Beckons Mama (B). The "O" Posture invites Dakota into a hug position (C).*

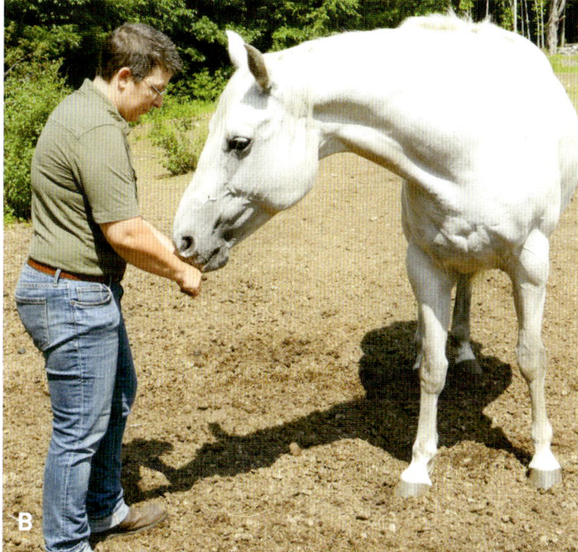



*4.9 A & B*
*I Beckon to Vati to come back (A), and she checks in with my Knuckle Touch and affectionately licks my hand (B).*

them. This is why it is so important to "talk" to them calmly and try to convey the fact that if they will allow us to use the Buttons wisely, including the fact that we are offering to be their Sentry and watch out for them, we will gain respect from deep inside the horse's very being—this is far different from the mechanical, external behavior learned through much traditional training and that only mimics the idea of respect.

## *Conversation:* Greet, Send Away, and Beckon Back with the "O" Posture

❶ Use your Knuckle Touch on the Greeting Button to just check in or say, "Hello."

❷ Use the Go Away Face Button with the Mid-Neck or Shoulder Button to move your horse's front feet over.

❸ Beckon your horse back with the "O" Posture. Remember, Beckoning welcomes the front end to return to us; it is a good idea to include a check-in Knuckle Touch as a Greeting when the horse responds (figs. 4.9 A & B).

## Reverse Gear and "X" Posture

When you want the horse to yield space by moving back, rather than just moving his head or front feet to the side, you can use the Button on the front of the horse's shoulder I call the *Back-Up Button* (see p. 39). The Button is low on the outside of

I had one student who was extremely excited about learning about her "X" and "O" and the profound effect they had on her mare. Her horse was standing nearby as my student was telling me about her experience in a very animated manner: As she talked, her posture become decidedly more of an "X." After a few minutes, her mare began dropping her head lower and lower, her ears flopping to the side as she gazed up at her owner's face. After the mare had blown a breath out her nose for the third time, I could stand it no longer, and pointed out that the mare was asking her human to *please calm down.* My student stopped mid-sentence, laughed heartily, and adopted a dramatic "O" posture. The mare instantly lifted her head to greet her owner's Knuckle Touch, lipping the woman's hand affectionately, as if to say, "There, there, dear, try to stay calm! There's a good girl!"

the shoulder, located where the very top of the leg comes into the shoulder muscle. There is an inverted triangular indentation here that can help you identify this Button. Touching it can make a horse flinch. Fighting stallions try to hit this Button with their front hooves.

## Core Energy

This is a good time to start exploring how sensitive your horse is to your *Core Energy.* Your true "center," your balance point, is located behind your belly button. This place in the human body is called the *Hara* in martial arts, which believes that energy broadcasts from your core. If we assume the horse has similar Core Energy, we should note that it is expressed through his chest as it radiates through from the depths of the center of *his* body (fig. 4.10). It doesn't matter whether you believe there is an "energy expression" from your core or not—you can still observe the horse's response to it every day.

We've already learned that "O" is a Beckoning Posture. Now it is time to learn about its counterpart: "X." When you face a horse directly—eyes, shoulders, hands, belly button, and feet—you are in the beginning formation of the "X"

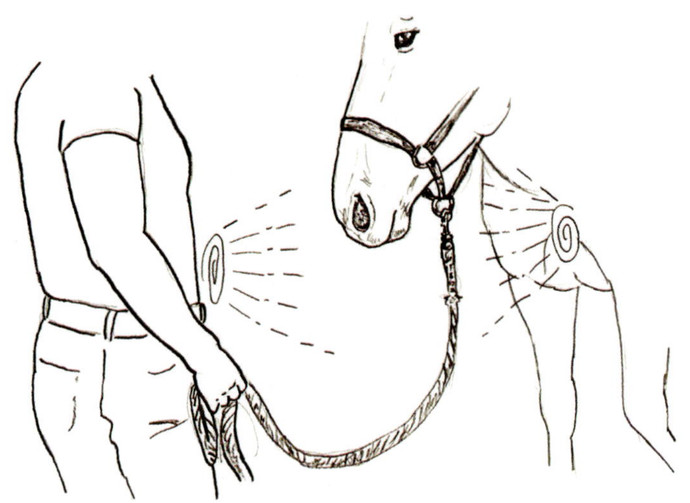

**4.10** *Your Core Energy radiates from your center just behind your belly button, and your horse's beams through his chest from his center body.*

Posture. Raise your hands or spread your feet and you begin to make a big "X" with its center at your belly button and your Core Energy (fig. 4.11).

Making an "X" communicates, "Go away." When a horse flattens his ears, arches his poll, spreads his front feet or hind feet or both, and seems to get taller, he is making an "X." His posture is not welcoming—it is not Beckoning you toward him. When you make your own body into an "X," you are mirroring this communication.

## Conversation: **Back Up**

I like to do this in the horse's stall to begin, and only later try it outside.

**4.11** *I make a large "X" shape with my hands up and my Core Energy aimed at Vati. This tells her to move away.*

❶ Most of the time just pointing your Core Energy at the Back-Up Button (aim your belly button at it) is enough to move a horse backward. You will find you really don't have to muscle most horses around to make them back up. There are some stoic horses that have shut down to people, and with them you can reach out with your hand to actually tickle or scratch this Button (figs. 4.12 A & B).

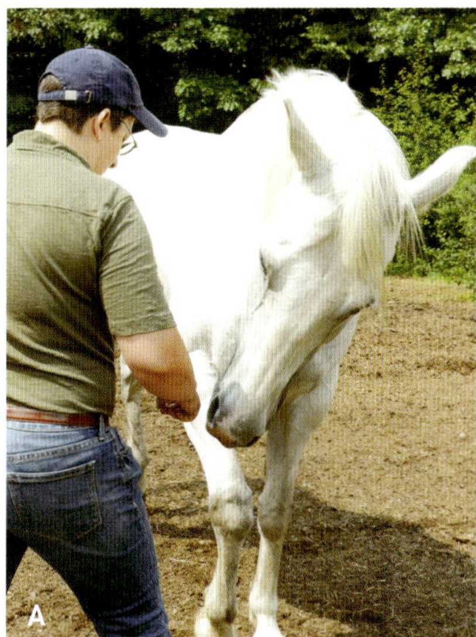

**4.12 A–C** *You can clearly see where Vati's leg meets her shoulder (A). I am pointing my Core Energy at it. To emphasize, I then aim my pointer finger at the Back-Up Button (B). I reward Vati with a Pause at Zero with my "O" (C).*

❷ Some horses that won't back up don't want to give you the space in front of them because they are questioning your rank as leader in the same way they would question another horse. When you ask for a Back-Up with these horses, it is *not* a matter of how far they back up. Praise their effort even if they just lean or show intention to move, thus creating success you can build on later. As we discussed in Step 1 (p. 12), learning to observe your horse's micromovements will alert you to many things, including the moment when he changes from "braced" to "thoughtful." Rewarding for the thoughtfulness encourages him to realize you want him thinking, not reacting (fig. 4.12 C).

### *Conversation:* Blocking Forward Movement

Blocking Forward Movement is related to the Back-Up. How? For the horse that wants to move forward, *staying in place equates to backing up in his mind*. I'm sure most of us have known a horse that liked to push past when his stall door or gate was opened. This kind of behavior can be changed if you talk to him about it in his language.

❶ To Block Forward Movement on the ground—that is, slow down or stop a horse you are handling or leading—stand slightly in front of and to one side of the horse with relaxed posture. Imagine your breath flowing down your legs and into the ground through your feet.

❷ Face the same direction the horse is moving. Pause slightly and turn your Core Energy as if shining a flashlight beaming from your belly button across the space in front of the horse like a barrier. See how he responds. If necessary, aim your Core Energy toward the Back-Up Button at the front of your horse's shoulder. You are claiming the space *in front* of your horse. A sensitive horse will not only Pause when you do this—slowing or halting—but will begin to back up if you keep your Core Energy focused toward his chest.

All the Conversations I present in this book fulfill the basic needs of horses with respect to what they consider important in their world. These Conversations let the horses know that what matters to them also matters to us. As a result, they might totally change their minds about what we are asking.

# THE CONVERSATION BEGINS

MOST OF YOU have horses that have had at least one previous owner. Daily, you are presented with habits, attitudes, and preferences based on prior history or conditioning. As you learn to interpret and mimic their language, you can unravel some of the mysteries your horses present. Come with me, in your imagination, as I meet and apply the conversational skills of Horse Speak with a big bay, off-the-track Thoroughbred. Let's call him Joe.

Joe is a study in contrast. He is physically talented and seems to like people—he's a pretty horse and not mean. But at times Joe bolts under saddle, weaves in his stall, and can be hard to catch. He can be friendly but other times, is defensive. He is often a challenge to lead and is nervous about doorways. His owners, Mike and Liz, are calm, considerate people who want to use Joe for light ring work and perhaps trail riding.

Other professionals have already tried to help Mike and Liz by overlaying training systems onto Joe's inconsistent behavior. However, I'm not here to train Joe, I'm here to have Conversations with him. I want to find out what he thinks about himself, what he understands about his environment, and what he feels about people.

My approach begins with elements of the Greeting Ritual, which is the set of gestures that horses perform in different combinations when they first meet or are turned out together in the morning (see p. 52). Using this Conversation, I will find out whether Joe's distress is linked to his mental reasoning, his emotional self, or possible physical limitations.

As I walk into Joe's barn, I become still on the inside (Inner Zero—see p. 7). I glimpse him in a stall at the far end and begin to observe him as I walk down the aisle. Horses have great distance vision so I can be sure he is already aware of my presence. The people around me are trying to keep up a friendly banter, and explaining what they think his problems are. Since horses live only in the moment, I keep my focus in the present moment as well. My attention leaves the world of humans as I focus completely on Joe.

As I approach, I watch his face for clues to several things: What does he think he needs to present to people? What is he actually feeling about being approached? Does he expect a treat? Or to be bullied? At what moment am I too close to his personal Bubble of safety... if ever?

I interpret all this information by observing the subtle cues of Joe's focus and attention on me. He is watching me with alert eyes and tense ears. His mouth is tight and his head is up high. He is not breathing much, but he is flaring his nostrils every now and then as if to get a sniff of my scent, which tells me he is curious. So far, Joe is accepting my approach with interest. However, his tense ears and raised head tell me he is already on alert—and ready for danger. Joe keeps looking at me with his head outside the stall—a more hopeful position than if he was hiding in the back of the stall.

Joe's ears occasionally twitch, an indication he reacts quickly. His tense face shows he is sensitive to a person's moods. I suspect his tight lips and clenched jaw point to the fact that he anticipates people might be in a bad mood around him, although he doesn't know why. He suddenly moves his head away from me and puts his ears backward.

> **The only agenda a horse has is whatever is happening in the current moment.**

I stop, which lets him know I understand what his gesture meant, that I am at the edge of his Bubble of Personal Space (see p. 28).

No one has ever stopped approaching him when he has asked before, and so this gets his attention. If a horse looks away when you are not close to him, it can mean he is feeling "invaded." Joe told me I had reached the edge of his Bubble of Personal Space—essentially his edge of safety where humans are concerned. All horses have a different size personal space, as well as humans. You can make a habit of noticing when your horse reacts, even slightly, as you approach. A reaction can happen when your Bubble touches his.

Now I am aware of Joe's boundary. I ask permission to continue to enter his space the same way that another horse would. I start by taking all the pressure off and turning away from him for a moment. I perform the human form of Aw-Shucks: scuffing the ground with my shoe a little and looking with interest at the floor (see p. 34). I even pick up a bit of hay and pretend to examine it closely. Horses do Aw-Shucks to take pressure off another horse or human by sniffing the ground. In his language, I am telling him I am in no hurry.

Doing the Aw-Shucks motion also gives me time to check in with *my* agenda— the pre-greeting Approach and Retreat that horses use with each other to keep calm and respectful. The only agenda *a horse* has is whatever is happening in the current moment. He has his wants and needs and will tell you directly what these are. I'm interested in Joe's messages.

Soon, Joe brings his head back toward me. This time he is sniffing more. He has lowered his head a fraction. He is showing he is curious about me. Although he is looking at me, I notice his eyelids are a little tightened. He is not yet convinced I have an honest and friendly intention, but he is open to talking about it.

Now the dance is on: I step forward one step, he withdraws his head to the side. I stop, look down, scuff the floor, and look in the same direction he just did. He turns back and looks at me. I repeat this sequence two more times when he looks away. Horses often have a pattern of three in their negotiations with each other. My question for Joe is, "May I approach you?" I normally repeat a question three times before I change my tactics if the horse doesn't answer me. I'm telling Joe it is his decision if and when I approach. His eye softens and his head lowers even more.

Finally, I am within reach of Joe. I officially request if I may join his Bubble by nodding my head and waiting. If Joe pins his ears, pulls away completely, or acts in any way rigid, his answer is "No." When I receive a "no," I scuff the floor and look away, then just sit down and breathe. I will wait patiently for him to allow me to connect. I am just following horse protocol: I have not been given permission, and I must wait for it.

Joe takes a break. He goes inside his stall and circles. He passes gas, snorts, checks for food and water, and appears to rudely tune me out. It is not rude, however, in the context of Horse Speak—he is thinking it over. I asked him if I could connect with him, in his own language. He is contemplating the fact that a person is talking to him in in a way he understands and he might be overwhelmed. If he has had a troubled past with humans, he is probably comparing my actions with his "database" of prior experience. After a few minutes, he puts his head out of the stall. His eyes are wrinkled and worried, and his lips tense, but he glances at me.

I stand again, turning toward him and holding my body at a slight 45-degree angle, which removes direct pressure to his head, then repeat my formal request: I nod my head again and wait. This time, Joe does not look away. Bingo: I just got permission to come closer. I step toward him, holding my arm in a relaxed way with my hand in a soft fist, fingers pointing down, and knuckles toward his muzzle. This gesture mimics how a horse's muzzle touches or reaches toward another horse's nose. My Knuckle Touch must be gentle because a horse's nose is sensitive.

There are normally three Knuckle Touches in the Greeting Ritual.

When I offer Joe my Knuckle Touch, I am on alert. It would be against horse protocol for him to bite me in this formal Greeting. However, there is always the chance he could offer me a defensive nip, which is a type of communication, as well. And, if he does, I am prepared to withdraw. Since Joe does not do this, I can keep the Conversation going. After the first Knuckle Touch, a horse will "play" Copycat, looking off to the right or left and waiting to see if you will, too. With this gesture, a horse decides who is the follower and who is the leader because the one who initiates that look is offering to lead, the one who follows the look is saying he is okay to follow, for now.

I initiate Copycat with Joe to demonstrate I understand horse protocol: I offer to be his leader, swinging my head to look to one side, and waiting to see if Joe turns his head in the direction I am looking. If he does, he is open to following me. If he turns to look in the opposite direction, he is not yet open to considering me as the leader but is interested in continuing the Conversation. If he withdraws into his stall, no matter how slightly, he is too insecure with humans to play Copycat. There is no wrong reaction: I'm just gathering information at this point. What Joe does indicates what sort of opinion he has about people he doesn't know.

Joe does not play Copycat after the first Knuckle Touch, which is also perfectly okay. I turn toward him and softly extend my knuckles toward his muzzle again. After the second Knuckle Touch, I copy his movement. He looks toward the door of the barn so I swing to look in that direction, also. I want him to know I am on his side and willing to meet him halfway. In a moment he brings his head back toward me so I turn toward him as well.

After this second touch and Copycat he pulls away, respectfully declining contact. He is not being rude, just politely withdrawing.

Since I have followed the Greeting Ritual protocol with Joe, and I see that Joe is still tense, his lips tight, eyes hard and wrinkled, and never dropping his head, I am prepared to be sent away but hopeful he will complete the ritual with a third touch. The third Knuckle Touch is the final one for this part of the Greeting. It determines the next step in the Conversation and requires sensitivity. The third Knuckle Touch leads into more (or less) connection and Conversation depending on the horse's needs and comfort level. In a herd situation, this part of the greeting can result in Grooming, erupt into play, or the two horses might agree to Share Space, moving off together in tandem or standing still in deep rest. Alternatively, it can result in one horse aggressively sending the other horse away with his Go Away language: a squeal, pinned ears, even a nip to the neck.

Basically, the third Knuckle Touch determines if you are friend or foe. Joe does not drive me away. In response, I offer to Groom him by slowly reaching back toward his withers, which is the protocol for affection. If he backs away, his answer is "No" to Grooming. (Grooming is Sharing Space. It is the horse allowing you into his Bubble of Personal Space.) If the horse seems engaged somewhat, then I know he *might* want to welcome me into his space—this is Joe's response. At first he is open to my Grooming gesture, allowing my touch, but then he leans away, not yet sure he wants me that close.

I now choose to demonstrate to Joe I know about personal space, both his and mine. The best way I can define our Bubbles is by asking him to give me space the way his mom would have. I reach up to the back of his cheek, to the Go Away Face Button (see p. 37) that says, "Move your head out of my space." I press it gently with the fingertips on one hand to clearly ask him to move his head away from my Bubble.

When he moves his head away, I step away and Pause, giving him time to absorb this information. I breathe deep and relax my jaw. I wiggle my lips, which means, "I am having a nice day, are you?" I've initiated this Conversation about personal space but also wish to keep our connection soft.

I tell a joke to the people watching. I make a soft but silly noise as I wiggle my lips. My sense of humor and deep breathing is crucial for communicating to him that I do not want our Conversation to be super-serious. I need to act like a peacemaker to a horse that is tense, unsure, and showing me in the way he holds his face that he has the potential to blow up. Why do I want Joe to believe I am a peacemaker? Because I never want to get into a fight with a horse! The actual occasions where I've had to keep myself safe with a truly aggressive horse are few and far between. I want to do everything in my power to convince

> I never want
> to get into
> a fight with
> a horse!

## THE CONVERSATION BEGINS

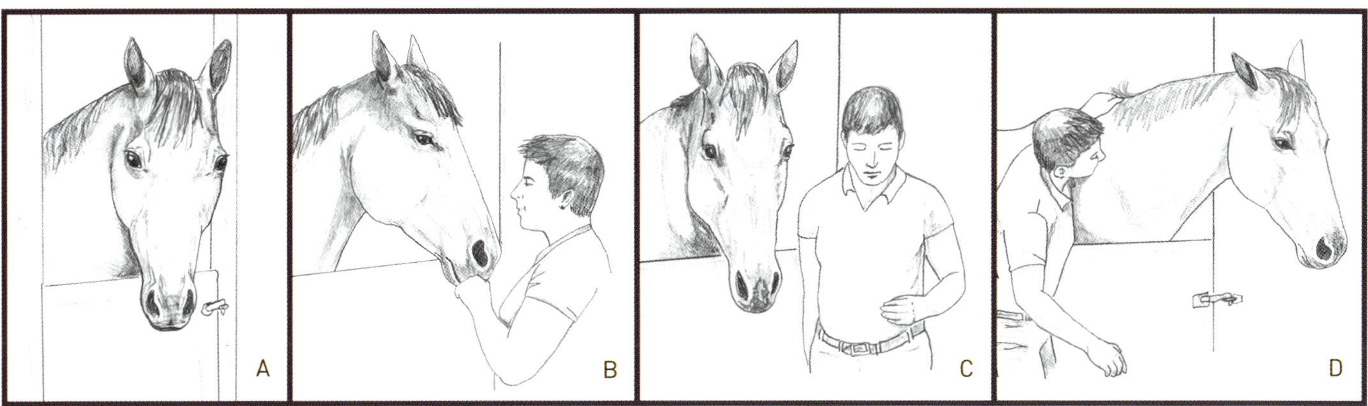

*Joe is alert, watching me come toward him (A). Our first greeting is hopeful but reserved (B). I practice Zero to take the pressure off (C). Then I move to Groom Joe's withers (D).*

a horse that I will do just the opposite of anything remotely resembling a fight or challenge. The most trustworthy leaders in a herd are also the calmest.

Since Joe has yielded to my space when he moved his head away, I now perform the Nurturing Breath (see p. 25). It is like a deep backward sniff, as though I need to clear my nose. The result is a sort of purring sound that dams make to their foals and I have heard other nurturing horses offer this sound to herd members in distress. I have been able to successfully use this sound to assure horses my intent is not only to listen, but talk and soothe as well.

At this point, I have:

**1** Completed all three steps of the Greeting Ritual (see p. 52).

**2** Asked Joe to move over by pressing the Go Away Face Button (see p. 37).

**3** Stepped away from his space, wiggled my lips, and offered the Nurturing Breath (see p. 25).

I've assured him that, while I do want to talk to him, I am not here to push him around; I am here to listen to him.

Now things get interesting. It seems as if Joe is the type of horse that always likes people even if he gets confused by them. Joe returns and wants to say "Hello" again. One more Knuckle Touch confirms that now I can use just one Knuckle Touch as a "check-in strategy." Joe lingers on my knuckles this time, and starts to lip them as though he is Grooming me. However, Joe is overexcited, and does not know how to do this softly and in a manner appropriate to my Bubble of safety. I must now use his good intentions to show him what kind of Bubble I want. I move his head away one more time, firmly but without malice. I reach to his neck and shoulder to scratch him. I also offer *Cupping*—a sort of patting in which I cup my hand and firmly pet him up and down on what I can reach of the top of his neck. This creates a relaxing feeling for a tense horse that wants to connect. If my touch is too light or unsure, a tense horse might feel irritated. I match my Cupping to his level of intensity, slowly encouraging him to relax.



Since Joe has lowered his head and relaxed a bit, I reward him by leaving. The only thing horses value more than food and water is space. Peacefulness in a herd happens when horses have their personal space a comfortable distance from another horse: space equals peace to a horse, so by giving Joe his space, I have offered him peace as a reward.

After 15 minutes, I return to visit him. I wait at the edge of the barn to see what sort of face he offers me. This time, he looks directly at me, without tension. He blows through his nose and Licks and Chews. His mouth is now relaxed. A horse that can Lick and Chew is feeling safe—he is literally "chewing it over" in his mind (see p. 42).

He is also Blinking, which tells me he is not holding tension in his eyes (see p. 45). His Blinking, in conjunction with his Licking and Chewing, indicates Joe has thought about what I have done in our last Conversation. He is interested in more dialogue.

I Blink at him, pretend to chew gum, and nod my head. I am saying, "Yes, that was quite a lot!" This tells Joe I am thinking about it, too, and would like to talk some more. He bobs his head slightly, and looks away for a second. He is checking to see if I understand that even though he wants me to come over, he is still unsure about our space Bubbles. I Copycat his movement, which tells him that I'm still paying attention to his signals.

Next, I nod my head to say I am coming over. I extend my arm, knuckles up, and walk directly to him. We do not need to do the long three-part, formal greeting this time because once is enough to slowly learn about each other, so I gently Knuckle Touch his nose three times in a row pretty quickly, but linger on the third one to see if he wants Grooming. This time, he lingers and wiggles his lips on my knuckles. I reach out with my pointy finger toward his cheek to the Go Away Face Button. I can maintain my Bubble of Personal Space because I can ask him to move his face over at any moment if I feel invaded. I reach up to scratch his neck.

When he overrides my hand and reaches to sniff my face, I am not surprised. Some of Joe's problems stem from being overly friendly and not knowing how to have boundaries with people because he does not talk human as well as other horses do. With Conversation, I can explain to Joe what people want in his own language.

I stop my affectionate scratching for a moment. In Horse Speak, I ask Joe to move his head one more time. I gently but firmly move his head on the Go Away Face Button. The reason? I am being a leader in determining the way our Bubbles of Personal Space interact, which continues to assure Joe: 1) I know about space Bubbles; 2) I know about invasion of space; and 3) I can tell him politely *how* and *when* to give me space.

This time, I do not leave. I step back, regroup, and allow him to regroup as well. When I offer my knuckles again, he predictably reaches for me. A lightbulb is going off in his head. He is beginning to realize I can read his questions and respond in his own language. And so he now asks me a very important question.

Joe is wondering about the dynamic of our relationship, just as he would if he was in a wild herd. Joe looks off at a real or imagined threat in the distance. I suspected this question might come up: if he is going to accept me as a peacemaker and leader, it has to include the role of protector, as well. In a herd, I call this protector role the *Sentry*— that is, the horse that makes the decision about whether or not the herd should run (see p. 49). This way, there are not 20 horses all making up their own minds whether or not to flee.

I look toward whatever Joe is looking at and make myself alert, fixated, and tense for a moment. I blow hard at the threat. Then I visibly relax. I drop my head, pretend to chew gum, and even take a big breath. Instantly, Joe drops his head and copies me.

I have said to Joe, "Yes, I know all about the 'bogeyman!' Don't worry, I will protect you!" Joe's mother was responsible for keeping him safe as a baby. She would have lifted her head, stared fixedly at a potential threat and would have blown at it loudly through her nose. Baby Joe would have stopped, watched his mother, and moved if she said so. When she relaxed, Licked and Chewed, then Joe could relax, too.

Joe has now found a way to initiate Conversation with me. Several times, he gets nervous about something. He is eager to see if I will pay attention to his fears every single time or not. I have to stop and look at his bogeyman every single time he sees one. I have to act like a protective mother if I want to win over his trust.

"Blowing the bogeyman away" does more to gain a horse's trust than anything else you can possibly do. It also sets you up for all sorts of wonderful desensitization later on. When he inherently trusts you to understand his fears and respond to them, he will begin to let you handle all the big decisions. Eventually, he will be less spooky and look directly to you when something makes him nervous.

Joe is beginning to trust me. I Greeted him with Horse Speak, I Groomed him starting at the withers like a horse, and I moved his face over like his momma did. I demonstrated the Bubble of safety I expected from him and blew the bogeyman away several times. I provided him many moments to stand and contemplate. I have let him have the freedom of his stall without a halter or lead rope on him. Joe is Yawning now, a sure sign he is letting go of years of confusion and tension.

# Horse Speak In-Hand

$P$erhaps you're beginning to feel like a field biologist, finding your horse the most interesting creature you've ever seen as you study every movement in detail. You've now practiced the Greeting when you "nod" at your horse, does he nod back? Do you hold out your fist in a Knuckle Touch as you approach your horse? Does your horse ever check in with you, reminding you to greet him? Have you seen your horse do Aw-Shucks, looking at the ground with great interest? Maybe you are trying to mirror him when no one is looking. Things are only going to get more interesting as you practice additional Conversations.

Up until this point, all of the Conversations I've outlined are intended to be done when the horse is loose in his stall, paddock, or pasture. Before we go any further, we need to discuss how to have *In-Hand Conversations*, so you can have the most success when handling and leading your horse with a halter and lead rope.

## CALMING HALTER CONVERSATIONS

By putting on a halter and lead rope, you take away the horse's ability to move independently. He may feel trapped. You assume a higher role in the herd hierarchy, even though he might still have his doubts about you being higher in the pecking order because previous humans may not have always been looking out for his best interests. You are asking him to

## Keys to Horse Speak: Step 5

**Rock the Baby with Halter (p. 79)**
**Rope Slide (p. 80)**
**Therapy Back-Up (p. 81)**
**Matching Steps (p. 85)**
**Target Hand (p. 87)**
**Fun with Feet (p. 88)**

follow you when asked, and he may have previous experiences being led that were negative. Simply putting a halter on can raise a horse's anxiety level. There are a few, short Conversations that encourage the horse to return to Zero after you put on a halter. One is a variation on Rock the Baby with Halter (see p. 55 for earlier discussion at liberty)—a great Conversation to have with a horse that is concerned with a light hold on his halter. The other, the Rope Slide, realigns your connection through the lead rope.

## *Conversation:* **Rock the Baby with Halter**

❶ Stand in front of your horse, facing him.

❷ With your left hand lightly holding the lead rope, place the fingers of the right hand palm down on the halter's noseband. Note: Some horses prefer your left hand to hold the lead rope a foot or so from the clasp instead of right on the halter. Do not put your hand too far around the noseband; just gently secure your fingertips there.

Immediately, shift your weight back and forth, first to one foot and then the other. Your hips can sway slightly. Breathe in as you rock in one direction and breathe out when you rock in the other (fig. 5.1).

When the horse relaxes, even a little bit, let go, and step away, softly using your voice in praise. Some horses that have been twitched or otherwise had their faces manhandled may resist Rock the Baby with Halter. You must breathe deeply with these horses and praise any release of tension or movement. Encourage them to trust your contact through Approach and Retreat, alternately touching and then letting go of the halter.

Next, step to the side of your horse's neck, with your left hand on the halter and your right hand either by your side or holding the rope on the withers to extend the rocking

*5.1  I do Rock the Baby with Halter from the front of Rocky.*

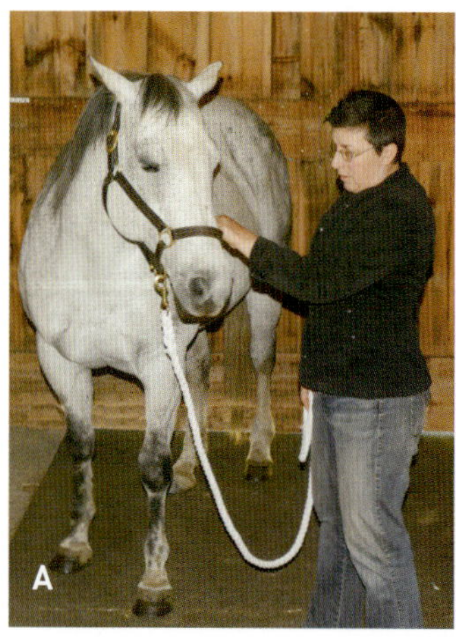
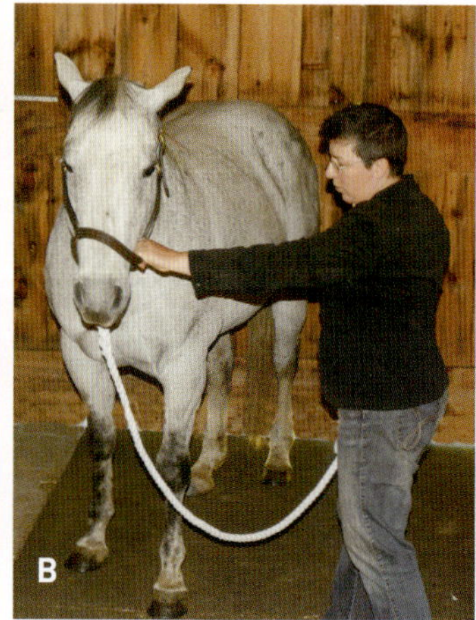
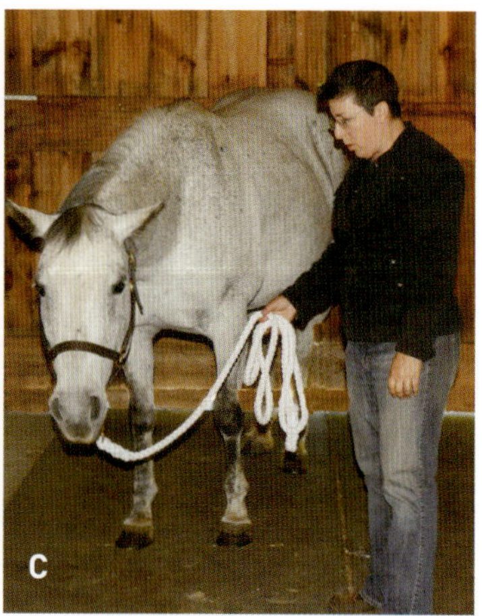

*5.2 A–C With Image, I stand to the side with my left hand on the halter and my right hand by my side (A). I am not "pushing" her head back and forth but instead creating a gentle rocking sensation (B). Image finds Rock the Baby very soothing (C).*

motion to the horse's body (figs. 5.2 A–C). Breathe in while rocking in one direction and breathe out when rocking in the other. When you sense resistance, try to rock through it for a minute, and then Pause before repeating the Conversation as often as needed. Eventually, the horse will realize you are offering a pleasant experience.

## Conversation: Rope Slide

**1** Stand in front of your horse and just to the side of his nose.

**2** At Inner Zero, hold the gathered lead rope with your non-dominant hand.

**3** Place your dominant hand's palm on the lead rope near the top and slide it all the way up to the clip.

**4** Your pinky finger is now toward the sky and your elbow is raised (fig. 5.3). Holding the rope palm down creates a resistance-free sensation for the horse and a fluid alignment from your arm to his chin. I believe this results in a vibration that soothes the horse.

*5.3 Practice sliding your dominant hand toward the horse to improve your light feel on the lead rope.*

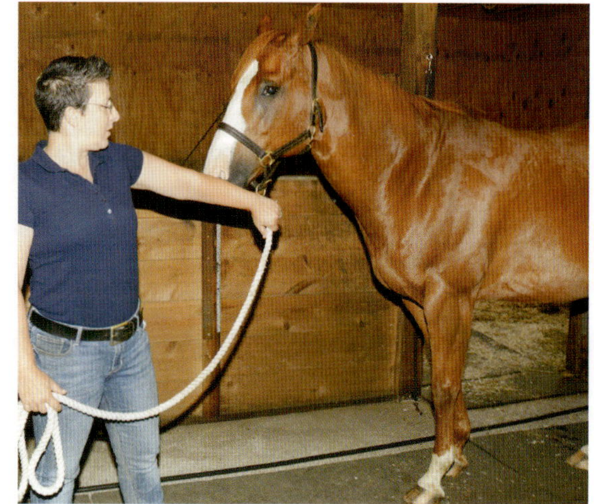

**5** No matter how your horse reacts, praise him and return to Zero.

Repeat the Rope Slide as often as needed when handling and leading the horse. It provides a comforting touch to add to your rope-handling skills.

## FURTHERING OUR CLAIMS TO SPACE IN FRONT OF THE HORSE

The Therapy Back-Up is the next Conversation to have with the horse's forehand. This Conversation is essential in claiming the space in front of the horse. We've broached this kind of communication already at liberty back in Step 4 (see p. 68). There we learned to use our Core Energy and the Back-Up Button to claim the space in front of him. To review: In Horse Speak, a leader will claim the space in front of another horse using first intention and then his eyes, his head, and his neck in sequence *before* he moves his feet to cause the other horse to Back Up and out of his Bubble of Personal Space. This communication requires that lower members in the pecking order stay on their toes in regards to observing the Bubbles all around them and how they connect. They need to be quick to yield the space in front of their heads to other horses. If the herd needs to run for cover suddenly, the horses do not want to trip over each other. In practicing spatial awareness in small doses all day, they are less likely to get hurt when under duress. It is so important for young horses to learn to back away from their elders that I have even observed older horses using their necks and shoulders to *physically* bump an inexperienced herdmember backward.

As we learned earlier, this amazing Conversation enables us to better control the space around the horse when we are near him or leading him. It is also valuable during riding. If you psychologically "own" the space in front of your horse's head, you will have an easier time guiding his movements because he believes in your ability to do so—even when you are up on his back.

### *Conversation:* Therapy Back-Up

Using a halter and lead rope, you will create a movement in your arm and "wing bone" (scapula) that is derived from Tai Chi and causes the horse to not only desire to move backward but relax his entire topline and release tension from his lower back, hips, and stifles, rounding his poll and using his core muscles, all at the same time. This is why it is *therapeutic*. Any movement a horse makes that causes him to feel pleasure he will memorize, and he will seek to repeat that movement. In addition, he will associate your request with a leader's ability to have good ideas.

❶ Face toward your horse's tail on one side of his body (not directly in front of him). Whether you begin this movement on the right or left is up to you. Most of us were taught to handle horses on the left side, so that is fine to start with (fig. 5.4 A).

**2** Place your hand closest to the horse's body, palm down on the lead rope, then slide your hand toward the clip at the end of the lead where it attaches to the halter ring under his chin (fig. 5.4 B). When done correctly, your pinky finger can feel his chin, and your elbow is slightly toward the sky.

**3** Your other hand, which should be holding the bulk of the lead rope, now moves towards the point of the horse's shoulder, aiming at the Back-Up Button—but only point at it.

**4** The hand holding the clip and ring of the halter now begins to rotate so that your thumb eventually faces upward with your elbow down. Engage this rotation all the way to your wing bone and into your back (fig. 5.4 C). This must be a slow, gentle rotation because it wields a surprising amount of power.

**5** Draw the lead rope toward the horse's chest, so his head and poll are encouraged to drop and arch downward toward the chest (fig. 5.4 D).

**6** Finally, take a firm, clear step toward the horse's front foot while using the appropriate amount of pressure on the lead for the horse you are working with. A very sensitive horse will move off at the first feel of the rotation of your arm on the lead rope (see Step 4). A stoic or stubborn horse may need you to use a significant amount of pressure. If the horse is "sticky," use your pointer finger to press into the Back-Up Button on the point of the shoulder where the shoulder meets the front leg. You only need ONE step. Insecure horses may only need to *lean* backward for you to be satisfied. Keep your eyes on the hind end of your horse—*not* his face, chest, or neck (figs. 5.4 E & F).

**7** The moment the horse yields backward even a little bit, return to Zero and praise him (and yourself!). Make the "O" Posture and Beckon him to return to your outstretched fist for a Knuckle Touch check-in.

When you request that the horse steps backward out of your Bubble of Personal Space in this way, you are saying you would like him to release his worries, tension, and stress. You are being a leader because you offer to create a safe space in both your worlds and always then invite him to please come back to your knuckles.

You can now use Buttons on the horse to ask him to:

- Move his face over.
- Move his front feet over.

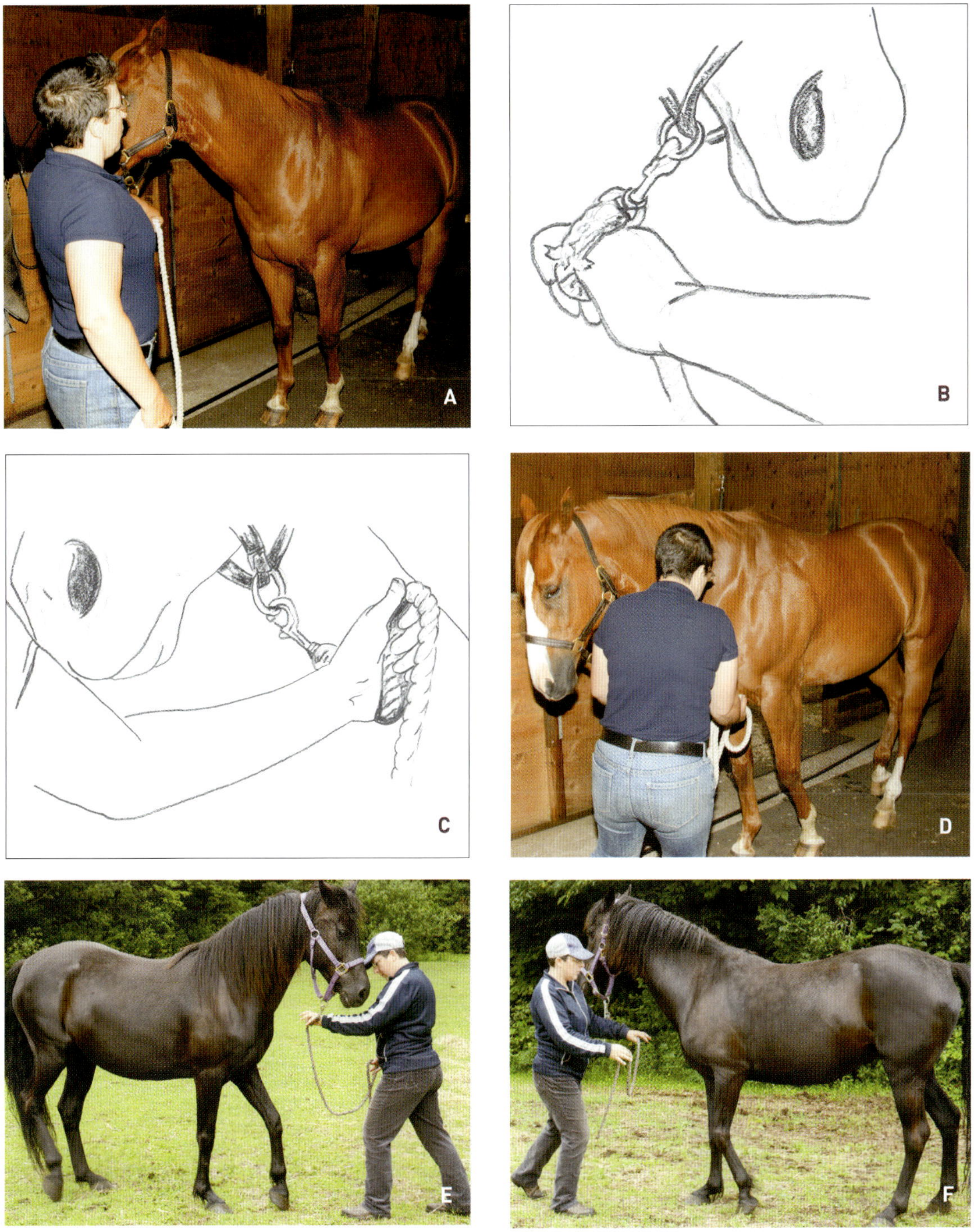

*5.4 A–F Begin the Therapy Back-Up, with all your body parts completely on one side of the horse (A). Your pinky slides up the lead rope toward the horse's chin (B). Rotate your entire arm so that your thumb turns up, creating a light feel that encourages resistance-free movement in the horse (C). I ask Clark to drop his head and arch at the poll. By exaggerating my "sit" posture, I also ask him to sink his haunches lower (D). Once you have the feel of how to rotate your arm, you will create wonderful softness in the horse (E). Pay attention to your horse's smallest try. One step back is plenty (F).*

■ Move backward out of your Bubble.

And you can step away from him and Beckon him to come to you, using your knuckles as a point of greeting.

## GOING SOMEWHERE PART THREE

### Moving Forward Together

Let's advance to the next step of the ritual of Going Somewhere—the point where you actually *do!* By now, you can Greet, find Zero, maintain a "How curious" attitude, and use Buttons to ask the horse to yield his front end away, and you can also Beckon him to return using "O" Posture. You have told him *you* are concerned about what *he* is concerned about in this world—namely lions, tigers, and bears (oh my), and you can use Horse Speak to address his concerns about these things by Scanning the Horizon, blowing a Sentry Breath, and maintaining Zero whenever you are near him. What you have learned to say to him is, basically, that when you show up, everything is safe, happy, and interesting.

You have built *trust* through the Greeting Ritual and *respect* through moving his forehand over, backward, and forward toward you. You and your horse understand each other's Bubbles of Personal Space, and you are now interested in taking it to the next level.

In order to move forward together, you and your horse must assume roles of leader/follower in your herd of two. There are two ways for a leader to move a herd member:

**1** The herd member follows the leader.
**2** The herd member is driven forward from behind by the leader.

In Going Somewhere Part Three, we are only concerned with *following* a leader. A young horse will stick to his mother's flank and orient his nose near her girth area so that he can move in any direction she does without getting run over by her. Should she move into his Bubble, he can pivot quickly away, as long as he is in the right position. Foals keep their muzzles connected to this area like a target, and they listen for their mother's footfalls to keep in-step with her speed and changes of direction. This is the very reason why these huge animals allow themselves to be led by humans and thin pieces of rope: it is already a part of their social connection to stay targeted on a leader and follow her footfalls. A connection also happens between two horses' shoulders. We've all seen horses driven in pairs, in perfect stride with each other. I've seen horses of *really* different sizes, matching strides in the pasture and even in a show ring. Energetically, they are attached to each other at the shoulder. When you line up your shoulder to your horse's, it is instinctual for him to match your pace and move with you.

To begin Going Somewhere, start with a Knuckle Touch to check in. Scan the Horizon and blow a Sentry Breath to let your horse know you have taken a look around and all is well, so now it is time to go do something. With your horse in a halter and lead rope, you will be further interacting with his forehand to proceed with the art of Going Somewhere, but this does not necessarily mean you will move any further than the threshold of the horse's stall door in this phase of communication. If your horse does not lead well, take time to ask for his space in his stall, first without and then with the halter and lead rope. Teach him your Zero so he can find his own. Do the Rope Slide, Rock the Baby with the Halter, and the Therapeutic Back-Up (see pp. 55, 80, and 81). Ask him to yield his front feet and step backward. In the aisle, repeat all these Conversations.

As you leave the barn, have the same Conversations about space-yielding as you approach the door, near the door, and again just outside it. Ask your horse to back up as you move toward the paddock. Try, as best you can, to *step with him,* and he will begin to see your feet.

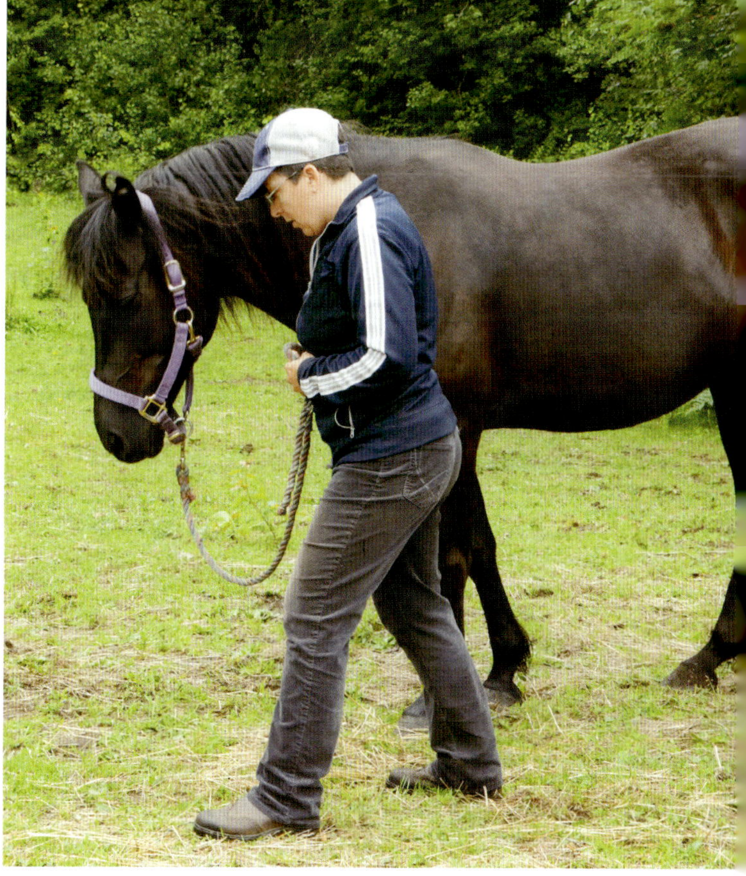

*5.5 Matching Steps can include mirroring other movement—here my head is low, like Mama's.*

## Conversation: Matching Steps

Every time you lead a horse you can tap into his sensitivity regarding hoof placement and rhythm. You can mirror your horse by matching your stride to his.

❶ Start by keeping pace with the rhythm of your horse's front feet. Have fun with Matching Steps: Take big, defined steps out of the stall, and over some logs or ground poles. Since we all have to lead horses to and fro, this makes a wonderful

Conversation while you are going somewhere together. Come into the moment and pay attention to his footfall, your head slightly lowered and in a relaxed frame—a walking Zero (fig. 5.5).

❷ When you can keep pace, ask him to start to change stride with you: Pretend you are in a marching band together, or *Stomp to a Stop,* and he will, too. The goal of being in step is not to have to use any intensity or pressure at all with your hands or arms when you are leading. Remember, it is natural for a horse to move *against* pressure so pulling on the horse will make him want to pull back. Also, most people just keep walking when they are leading. Your friendship is better served by Stomping to a Stop occasionally, asking for a Back-Up or sideways yield using the Buttons from earlier Conversations, and being in step with the horse.

Use Horse Speak *with your feet* to rate the speed while walking with your horse. What do I mean by this?

- Stomping both feet can bring him to stop.
- Strutting can lengthen his stride.
- Crouch and pretend you are walking on thin ice, just to see how your horse mimics you.
- Your horse will become so focused on your feet you can drift sideways and he will follow.
- Cross one foot over the other, and your horse might step across one front leg, as well.

By emphasizing the feet and deemphasizing the halter and lead rope, you may graduate to leading with only a neck rope, and eventually, you will be able to do it with just a piece of mane. Mane-leading is one benefit when you and your horse both practice Matching Steps (fig. 5.6).

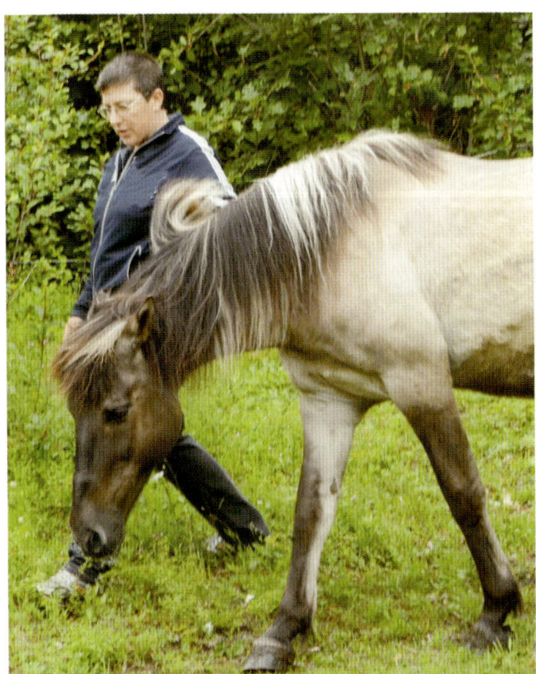

5.6 *Rocky and I are Matching Steps, and I can easily lead him with a piece of mane, nothing else.*

5.7 *Keeping my lead hand out for a check-in turns it into a target for her to follow.*

## Target Hand and Understanding the Horse's Circles and Arcs

Now, using your greeting Knuckle Touch as a consistent place to check in means your knuckles can also be used as a target the horse will easily follow. I find that horses generally love this, and many of them take an attitude of, "*That's* what you wanted all this time? Well, why didn't you just say so before?" You will find that you can allow some space on the lead rope if you hold your knuckles away from your body, up near the horse's face, allowing him to follow your hand as a target, rather than relying on the lead rope for guidance (figs. 5.7). The lead rope becomes superfluous and is merely present as a reminder to not wander off, or if need be, a safety device for emergency situations.

Remember how we talked about the horse moving in circles and arcs back on p. 28? One way to understand these better is to move in tandem with the horse as he follows your Target Hand. As you move forward, make sure you are Matching Steps, staying near his neck and shoulder. Then, make a gradual change of direction (fig. 5.8). If the horse does not follow you, your arc was too sharp (more like an angle) or too sudden. We humans tend to move in angles, not arcs, so this is not surprising.

To assist yourself in making better arcs and circles with your horse following your Target Hand, set up some traffic cones to give your eyes something to focus on and you a point to travel toward. Once you get good at having the horse follow your Target Hand, there are all sorts of neat things to try like stepping over ground poles, weaving in and out of objects together, or trying it without a halter and lead rope (fig. 5.9).

Again, when you ask for something and you get something, say, "Thank you," with a Pause at Zero. This is about *consistency*, not control, and *progress*, not perfection.

**5.8** *While Matching Steps I use my Target Hand so Mama will follow me in an arc as we change direction.*

**5.9** *When you get good with the Target Hand and Matching Steps, you can do arcs and circles without the halter and lead rope.*

## Variations for Fun

By now, you will have come to understand the world as your horse sees it, and he knows he can now use Horse Speak to reach out to you if he really needs your guidance, attention, or maybe he has a bright idea. You can use your Target Hand to guide your horse forward, and you can Beckon to him, back him up, move him side-ways, and Match Steps around and over minor obstacles like thresholds, so you are well on your way to what I call *Fun with Feet*. We are still focused on working with the horse's forehand, and by doing so we are developing more trust and respect with every Conversation we have.

## *Conversation:* Fun with Feet

You can take mirroring and Matching Steps to the next stage by asking for *just one step*. Horses adjust their space with each other by positioning and repositioning their hooves. And they understand *your* language of claiming and yielding space when you use your feet with clear, calibrated movements. Fun with Feet is the Con-versation that introduces this. You can face your horse or do this exercise from the side of his shoulder.

❶ Slowly and deliberately step out wide and watch your horse follow (fig. 5.10 A).

❷ Then cross your feet dramatically (fig. 5.10 B), followed by stepping wide again. Give your horse a chance to follow your lead. If he misses it the first time, start over.

❸ Now face your horse from the front, and step dramatically toward one of his front hooves. If you have already practiced the Therapy Back-Up (p. 81), you may use a light touch on the lead rope to help clarify what you want. However, if you have Matched Steps enough, your horse should be clued in to your footfalls and simply lift the one front hoof and shift it backward as you step forward. This needs to have the same "marching" feeling to it that Matching Steps had.

❹ Try some variations, such as stepping forward and backward with the same foot/hoof combination, or stepping forward two or three times in succession, and having him mirror you, then stepping backward two or three steps, while he brings the same hoof forward.

❺ For extra fun, stop mid-stride and dangle your foot in the air. I have seen horses wiggle their ears and lips in what I call a Horse Speak "giggle" (see p. 16) as they try to suspend their own hooves (fig. 5.10 C).

**5.10 A–D** *Begin Fun with Feet with a step out wide to the side (A). It may take three or four tries, going slowly to get the timing (B). "How long can you stand on one foot?" (C). When I Stomp to a Stop, Dakota lines her feet up with mine (D).*

**6** Stomp to a Stop and see if your horse mirrors you (fig. 5.10 D).

Now, when you start to do Matching Steps forward with your horse to Go Somewhere, you can change it up, and turn to face him, stepping in toward his hoof to ask for Fun with Feet. This keeps your horse on his toes and makes you seem very fun and interesting to him. On a practical level, you are gaining balance, harmony, rhythm, and a better feel for your horse's natural body movements. Plus, when you are riding, you will be more in tune with his feet, and this will add value to your mutual awareness.

### Conversation: Obstacle Course

With Obstacle Course work, you have a chance to work out how to Go Somewhere Together and have some fun while doing it.

Use whatever items you have at your disposal to engage all the Buttons for sidestepping, backing up, marching forward, and Beckoning to you. Set up scenarios in which you back your horse up in strange places, or Beckon him to come toward you over a ground pole or between some barrels. For safety purposes, always stand at a 45-degree angle to the front of your horse when incorporating obstacles so if he lacks confidence and needs to "scoot" over or forward, there is room for him to move without running you over. If you have claimed the space in front of him with the Therapy Back-Up as we've discussed, he will make every effort to be careful to avoid your Bubble of Personal Space.

❶ Matching Steps, marching forward is a real treat for horses to practice. Set up ground poles at odd angles (fig. 5.11). This will encourage you both to really Match Steps as you negotiate the pace and height you need to lift your knees to clear the rails.

5.11 *Matching Steps over ground poles gives you something fun to talk about!*

❷ Bring out balls, hang objects in corners, set up barrels, go up and down hills, linger around gates, doors, and anxiety-inducing objects. The point is simply to have lots of things to talk about! Figure out what brings out your horse's sense of humor, and yours, too. Also determine what makes your horse feel like he has done a good job: Some horses enjoy a good scratch, others like toys, and some only seem to have a light bulb go off when there is a cookie involved!

# Grooming Ritual: Finding Unity

Now that we can enjoy Going Somewhere in a variety of ways, let's revisit that other G we introduced back on p. 51: Grooming.

In Horse Speak, Grooming does not always include touching. For horses, *Social Grooming* simply means sharing affectionate energy on any level. Because, as we've discussed, horses value space more than contact, their favorite form of connection is a simple one.

## SHARING SPACE AND REVISITING PAUSING

Sharing Space means just that: two beings standing, sitting, or napping happily near each other in total Zero. This is lovely for horses but can seem a bit boring to us. This is why *Pausing,* which we first learned about on p. 23, is such an important prerequisite when learning Horse Speak.

Remembering to stay Zero encourages you to Pause more often, take a breath, and relax. Even though this is maybe not the equivalent of the long hours, dozing in the sunshine, that horses enjoy with each other, Pausing is a version of Sharing Space that we humans are more likely to remember or have the patience to do.

**Keys** to
Horse Speak: Step 6
**Sharing Space (p. 91)**
**Making Contact (p. 92)**
**Managing Your Belly Button**
 **(p. 95)**
**Drop It to Stop It (p. 98)**

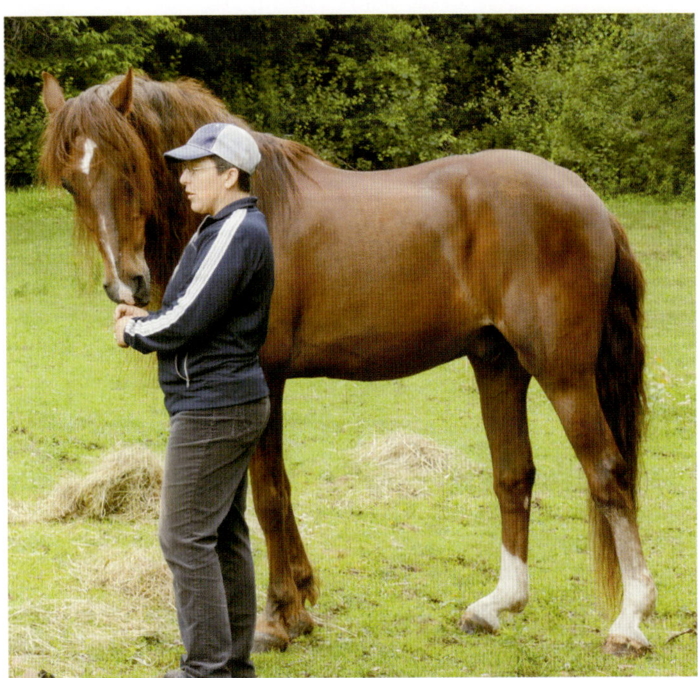

*6.1 Zeke checks in with my "O" Posture as I invite him to Share Space.*



## Conversation: Sharing Space

❶ Pause: This invites a Conversation about the day.

❷ If you wiggle your lips, add Aw-Shucks, take a deep breath, or assume "O" Posture, you say that you feel fine in the moment and would like to savor it. Horses *love* to savor quiet moments. You will probably find your horse dropping his head, sniffing the ground (the horse version of Aw-Shucks), flopping his ears to the side (part of the horse's "O" Posture), wiggling his lips in turn, and taking a deep breath (fig. 6.1).

For many horses, this conscientious and mindful Pause is a Sharing Space Conversation that will comfort them. This is part of Social Grooming in Horse Speak. If you Pause long enough, your horse may in fact offer a delicate Greeting to you in which he lips your knuckles—an invitation to touch and mutually groom each other.

## MAKING CONTACT

To touch a horse, I always start out on the upper neck or withers the way another horse would. In Horse Speak this says very clearly that I desire that same level of closeness. I hold out my knuckles toward his lips so that there is a place for him to lick me in response. Very rarely will a horse nibble the back of my hand or become a bit "too much" at this juncture, but if that happens, I simply use the Face Go Away Button to ask for space in my Bubble of Personal Space and do Aw-Shucks while maintaining my own scratching on their neck or withers (figs. 6.2 A & B).

When a horse is too high energy for a calm scratch, I may use *Cupping* on the neck (fig. 6.3). I first talked about this technique in "Joe: The Conversation Begins" (p. 76)—it is when I shape my hand into a "cup" and pat the horse in a rhythmic sequence down his topline. Cupping helps release strain in the fascia that attaches, stabilizes, encloses, and separates muscles. It also releases endorphins, which

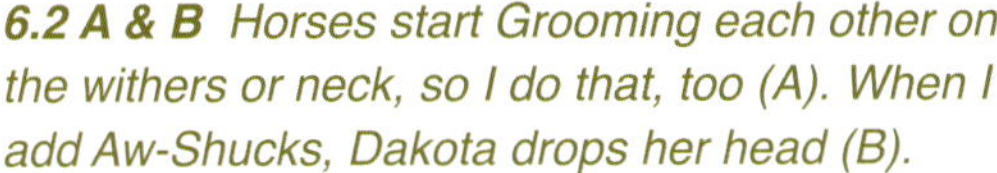
**6.2 A & B** *Horses start Grooming each other on the withers or neck, so I do that, too (A). When I add Aw-Shucks, Dakota drops her head (B).*

aid in relaxation, especially in tense horses. Reiki and Tellington TTouch® are two other fabulous kinds of bodywork that offer a means of connection.

Once I have made initial contact on the neck and withers, I move on to whatever Grooming I would like to do with my horse—whether it is purely emotional (scratching or stroking) or pulling out the brushes and getting some mud off. Either way, I have found that by starting my Grooming session from the "Horse Speak perspective"— Sharing Space first and then scratching the withers—my horses not only tolerate "technical" Grooming I want to get done, they positively lean into it.

For those horses that enjoy being brushed anyway, Pausing for a moment to Share Space only adds value and depth to the experience. Many horses, however, do not enjoy brushing or being touched, and stand tensely on the cross-ties or fidget the whole time. Sharing Space is a way to say to them in Horse Speak that you

*6.3 Cupping a tense horse on the neck can be very effective in helping him calm down.*

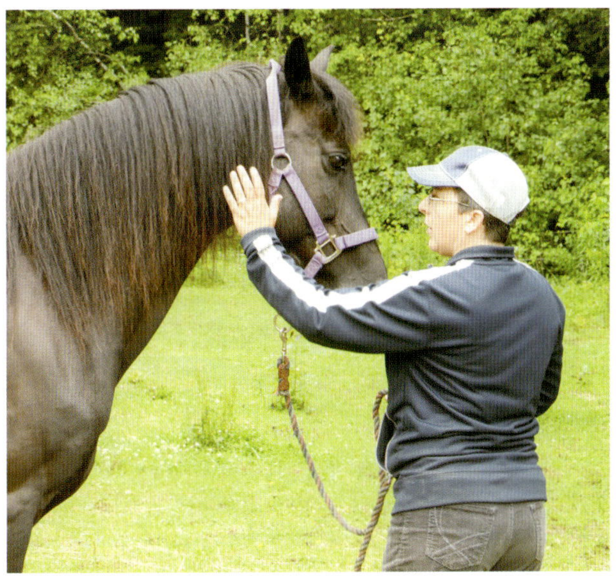

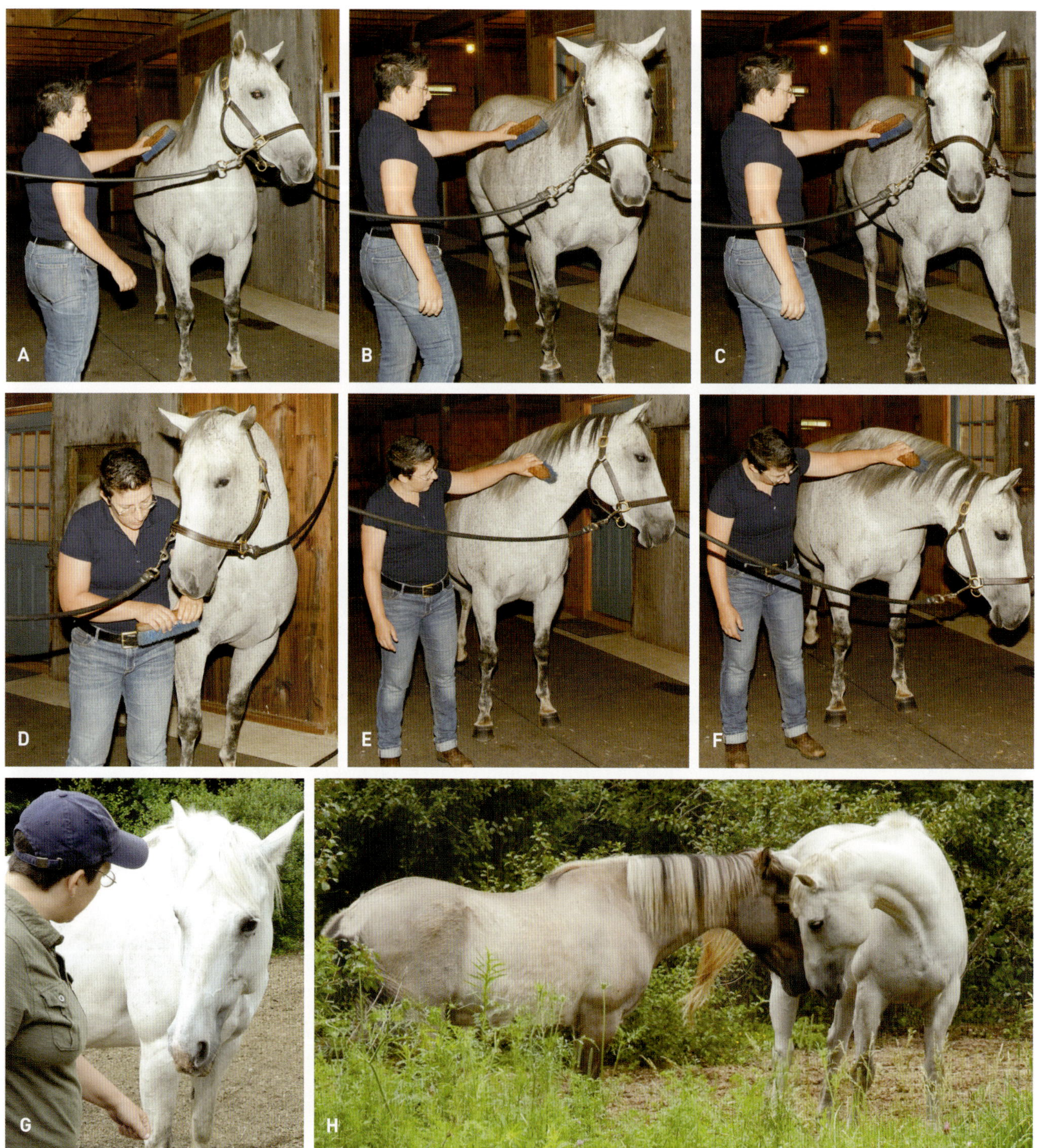

*6.4 A–C* When I face Image, she wants to yield her face away from me (A). My accidental "X" sends her mixed messages (B). She tries to shift over, away from the pressure she feels (C). I change to "O" and Image checks in (D). By keeping my belly button and its Core Energy off her and maintaining an "O," I help her relax (E). She defers space with her head, but also lowers it as her anxiety lessens (F). When I stand at a 45-degree angle to the horse, she can easily move away so doesn't feel trapped (G). Horses often use this 45-degree angle with each other to keep the pressure off (H).

would like to engage your Bubbles of Personal Space more directly and intimately, and would they be open to that?

## Don't Forget Your "Xs" and "Os"

Remember, horses are reading your body language—even if you are not fluent in it! To the horse, being on cross-ties while you face him with your belly button sends a confusing message. Remember what we learned on p. 68 about the power of your Core Energy and your "X" Posture: When you face a horse directly with your hands raised and your feet spread, you are saying "Go away." And when the horse is on the cross-ties, you are saying, "Go away...but you can't." You have to consider your stance and movements to prevent misunderstandings.

### *Conversation:* Managing Your Belly Button

As you groom your horse, your arms rise up, and your feet often spread out as your belly button faces him. You have just formed an "X" with your posture, even though you didn't mean to (figs. 6.4 A–C).

The answer is simple: Keep your belly button *off the horse.*

**1** Stand at what is about a 45-degree angle at Zero and watch his eyes, lips, chin, and ears (figs. 6.4 D–F). Look for signs of relaxation.

**2** Pay attention to your breathing and your horse's, too. Some horses hold their breath when you have your belly button on them, and if you are paying attention, you may realize you started holding your breath, too!

**3** Staying Zero will help you realize when you are slipping gears and becoming accidentally tense. Pay attention to your belly button in all that you do with your horse on the cross-ties, because it can be the secret cause of many misunderstandings. Maintain a 45-degree angle (6.4 G & H).

**4** Switch your posture to an "O" at any moment to demonstrate that you realized you were becoming an "X" and using your Core Energy wrong. Get in a habit of making an "O" whenever you or your horse feel tense for any reason. Your Outer Zero will keep your Inner Zero calm. These "X" and "O" Conversations are so key in Horse Speak! Your horse will start to find ways to talk to *you* about it.

# Sharing Space
## IN PRACTICE

While taking pictures for this book, I turned my tense, reactive, and hard-to-please mare Mama loose with my older, self-confident, calm gelding Zeke to try to capture a "real, live" horse Conversation in visual detail. What was captured was a pretty clear example of horse dialogue and the natural progression to Sharing Space (figs. 6.5 A–U).

Mama and Zeke Beckon to each other (A). Although they see each other all the time "over the fence," being alone together means they have to establish who's who and what's what in the "herd" immediately.

As Mama and Zeke approach each other, they keep their heads very low (B). Knowing Zeke, I would say that he is taking the lead, and Mama is playing Copycat. I say this because I know their temperaments, and I know that Mama typically does everything with a very high head. Zeke would have to tell her to stay calm as they approach each other.

Their first official Greeting is performed over a bite of hay (C). Zeke tells Mama to chill out, relax, and have some lunch. The first "official" Copycat is done with their heads low. However, Mama does not follow Zeke's lead perfectly: She moves her head in the opposite direction, already testing his leadership (D).

Now Zeke means business and raises his head to Greet Mama more seriously for the Second Touch. She swishes her tail, telling him she does not appreciate his boldness (E).

But Zeke's efforts worked (for the moment)—Mama Copycats him to the same side (F).

Before the Third Touch happens, Zeke decides to walk a small circle and think things over. He keeps his head low, which says there is not really a problem, but swishes his tail, telling Mama not to follow (G). As soon as he returns, however, and reaches into Mama's Bubble of Personal Space for the Third Touch to offer to connect and possibly comfort her, she dramatically swings her head away, Scanning the Horizon and swishing her tail. This says she does not feel safe and does not want to connect like that (H).

Before Zeke has time to address Mama's concerns, she takes off at a dead gallop (I). Mama is a bit of a "drama queen," and Zeke is a little surprised and annoyed that she took things to this level. His ears go back, his face shows irriation, and he swishes his tail disapprovingly. Catching up with her, he uses the Shoulder, Mid-Neck, and Go Away Face Buttons to push into Mama's Bubble and tell her to circle around him (J). He aims his front leg at her shoulder as he exaggerates his footfalls to try to get her to Match Steps with him and slow down. He is really mad: His ears are flat and his tail is *really* swishing.

Mama ignores his direction in a panic, running off even faster. Zeke, around 26 years old at the time of writing, is in no mood to chase her around.

He simply goes to the middle of the field and goes to Zero (K). He is saying he will just stand there quietly, waiting for her to calm down. When she is ready, they can talk again.

Knowing that Mama can really work herself up, I decide to step in. First, I approach Zeke, at Zero and in my "O" posture, to show her that he and I are bonded as leaders. Zeke Beckons to me, no doubt hoping I will help him with this crazy mare. As I get close to Zeke, Mama turns to look at me—but this does nothing to slow her down (L).

Staying close to the center of the paddock near Zeke, I start to Match Steps with Mama's stride, keeping my "O" Posture and slightly nodding my head to tell her she can come to me at any time (M). After a few strides, I use what I call *Drop It to Stop It:* I change my posture to look as though I am about to sit down (N). This tells Mama I

want her to stop, too. Since we were Matching Steps and she is connected to me, she eventually does try to honor my request. I ask her to come over to me the same way Zeke did in the beginning, with a low head, and equally low intensity. I round my shoulders and look down to really be clear that I want her to calm down (O).

I pick up her halter (which was just off to the side on the ground). Before I try to put it on, I Scan the Horizon very dramatically, since that was what set her off in the first place (P).

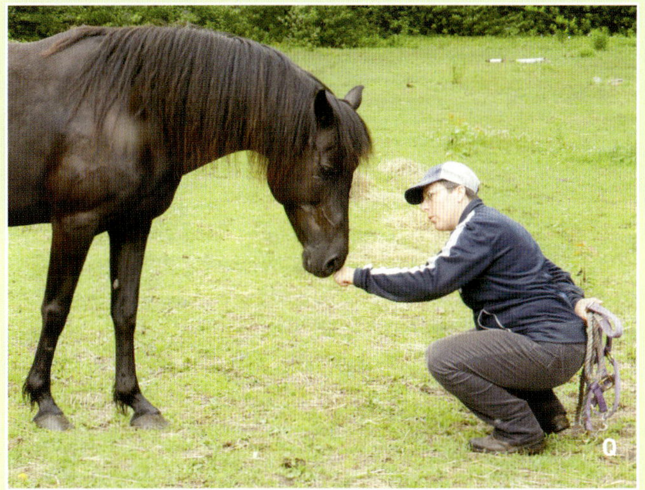

I do a check-in Greeting, dropping myself into a squat to ask her to truly be at Zero with me and extending my knuckles (Q). As I place the halter on Mama, I put my back against her shoulder, taking all pressure from my belly button and Core Energy off, and inviting her into a "hug" position (R). You will often see horses that are being "soft" with each other standing at an angle like this.

Now, we Share Space: I am at Zero, in my "0" Posture, with one hip cocked. We breathe deeply. Zeke walks off to go do something else now that the action has ceased (S). Then he decides to come back and check in, making sure all is well. I keep my head lowered to ask them both to stay calm near me (T). Then I indicate toward the Go Away Face Button on both horses, claiming the space and saying that I am everyone's leader. My head is still low, because I am not making a big deal about it. I want this interaction to stay at Zero (U). Now Zeke knows he can relax because I have taken charge.

# The Five Levels of Intensity

$A$s we discussed in brief in chapter 1 when I introduced "managing volume (see p. 9), I have assigned numbers to represent the levels of intensity of physical movement (or "volume") you might use in a Conversation with a horse. To recap, the five numbers are Zero, One, Two, Three, and Four. As the numbers increase so does the size or emphasis of your movement (figs. 7.1 A–F).

## LEARNING THE LEVELS

Now that you have some experience with some of the Horse Speak basics, you can learn to adjust your language intensity to correspond with a situation. You have learned the alphabet and your first sentences; now you can learn to speak and read with inflection and fluency.

### Outer Zero and "O" Posture

It all begins with complete calm in mind and body—Zero. We explored Inner and Outer Zero in detail on p. 7. Here we will simply revisit it so we have a place on the intensity spectrum from which to begin. As you know, the "O" Posture is Beckoning, and the inward curve of your body welcomes the horse to approach while in a state of Inner Zero (see fig. 7.1 B).

## Keys to Horse Speak: Step 7

 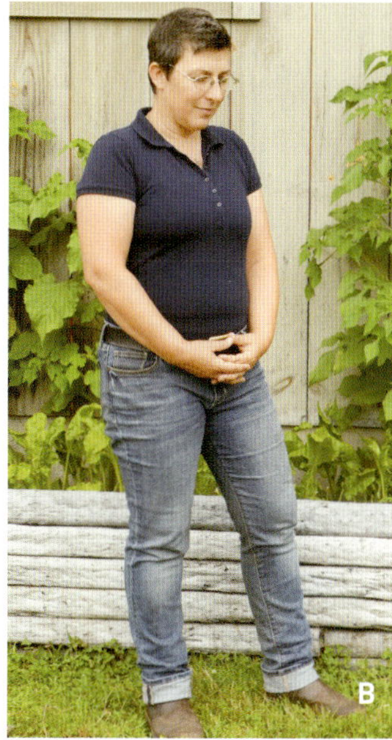 

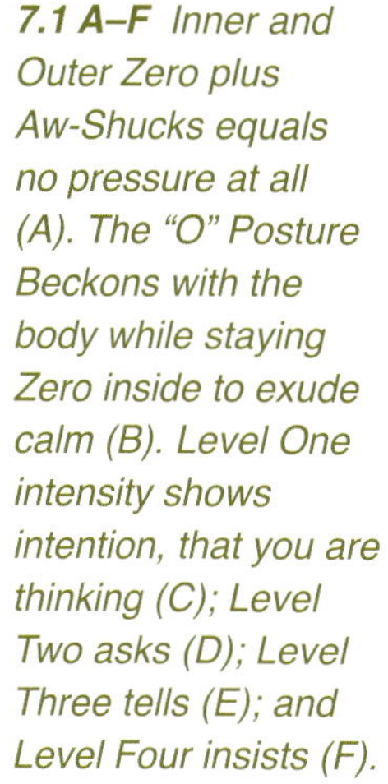

***7.1 A–F*** *Inner and Outer Zero plus Aw-Shucks equals no pressure at all (A). The "O" Posture Beckons with the body while staying Zero inside to exude calm (B). Level One intensity shows intention, that you are thinking (C); Level Two asks (D); Level Three tells (E); and Level Four insists (F).*

## Level One

The more common usage of the word *intention* doesn't seem to mesh with physical gestures, as it assumes a mental state rather than a physical one. They are not mutually exclusive, however. When you hold an intention in your mind (a determination to act in a certain way or do a certain thing) your body shifts ever so subtly. And guess what? Your horse can read *any* microscopic shift in your posture: the tilt of your chin, the angle of your shoulders, the bend of your knees. Therefore, the way to express Level One intensity is through your posture. With intention, you stand straighter than you do when you are at Inner and Outer Zero (see fig. 7.1 C). Your head is higher and your shoulders are back (rather than curved in as with the "O" Posture."

Sensitive horses respond to Level One intention all the time and you just don't notice it as they shift their bodies and yield their space to you when you turn to face them. Stoic horses, on the other hand, may not respond much to your upright posture energy. It can take a while for them to understand what you are doing. These are often horses that have "shut down" in order to cope with inconsistent communication from people. Lesson horses are usually pretty stoic from years of dealing with unbalanced riders and mixed messages. At first, you may need to use higher levels of intensity with stoic horses, but I have found they can be just as sensitive as any other horse once they feel they can open up to people in Conversation.

## *Conversation:* Practicing Posture

Explore your posture at random times with your horse. Say you are walking toward the paddock and your horse is watching you come toward him:

**1** With your emotions at Zero, square your shoulders to your horse.

**2** Then focus on the side of his neck and visualize him turning away. See what his reaction is.

**3** Return your body to Outer Zero and/or turn completely away from him.

Engaging my core with slightly bent knees and at the same time imagining my breath so it *seeps* into the ground has a number of times stopped a horse running straight at me. You have to be willing to play a bit with this as you learn.

Level One intensity can be dramatically quite effective even though you do not move your arms or legs in order to convey a message. You simply form an intention in your mind, look toward a Button on the horse, and straighten your posture so it begins to resemble the "X" you learned on p. 68 (and we'll go into more depth when it comes to this subject, beginning on p. 107). The important thing is you return to the relaxed posture of Zero as soon as you can.

## Level Two

Motion is now added in order to convey a message. Level Two is *asking politely*, combining Level One intensity and simply adding movement. Progression to Level Two looks like this: Your mind is relaxed at Zero, then, holds a clear intention of what you are asking the horse to do. You stand in a more upright posture and focus on the Buttons appropriate to your request. If your horse does not yield or respond, *add to your intention* by pointing to the Buttons with your fingers or waving your hand for direct motion. You may also extend the reach of this indirect pressure by waving or shaking a piece of baling twine or a handkerchief toward the Button, or a plastic bag tied to the end of a crop or stick. You can move a horse out of your space by gently flapping your hands in the air. I like to start with a "tickle motion" with my fingers toward Buttons—not only does it keep me less serious, the horses also seem to appreciate this style of gesture.

Level Two can be a pivot point where you'll begin to know whether you need to either "turn up your volume," or perhaps, tone it down. It is not uncommon, for example, for a horse to suddenly become non-compliant. This is because two horses will eventually explore *all* the levels of intensity with each other to see what each horse's levels look like. If one horse flies up the spectrum, resorting to force easily, then the other horse knows he should avoid making waves. Horses want to know what *your* intensity levels look like, for this same reason.

## Level Three

When you "turn up" the intensity another notch, it is more active, and motion of the hands is joined by a step *toward* the horse (see fig. 7.1 E). It is the equivalent of using a louder "physical voice," closer to "telling" in human inflection than "asking." Keeping the calm of Inner Zero emotionally, set your intention with your body posture, but this time, step *toward* the horse and motion to a Button. Move into his Bubble of Personal Space. Often the horse feels your Bubble reach his and reads your intent to touch him before you actually get close enough to do so. When your horse does not

**Intensity Level Distinctions IN PRACTICE**

We were just discussing how all horses have different sensitivity, and here is where you will see how important it is to gradually work up in levels of intensity until you are heard, then come back down to the lowest level at which you can still convey your message. For example, stepping toward a really sensitive horse with what you think is Level Three, might be the equivalent of using Level Four. You need to watch for a lean, twitch, or other subtle response in order to know how effectively you have communicated. *Any* answer by the horse is a good one, and that means you should go back to Zero. You really want him to understand your distinctions.

hear or read your intention and does not listen to motion, you can now thoughtfully add touch to the message. Incorporate just enough pressure to express your message, then stop and return to Zero. If you stay thoughtful, so will your horse.

Horses avoid actually biting and kicking the majority of the time—the equivalent of very loud touch or tell. With Level Three intensity, you risk creating resistance with such an expression. When horses create resistance in each other all the time, there isn't peace in the herd, which we already know they crave. So it is important for you to use *just enough* intensity to communicate your message. On the other hand, you have to hold your intensity level until the horse at least flinches an acknowledgment of you. The more tuned in you get to the language, the more subtle all of your interactions become with your horse.

## Level Four

Now is the time to demand, but you still are not "screaming" because no matter what your physical gesture is (now an obvious "X," spreading your legs apart and lifting your arms), you want to stay Zero *inside.* Level Four can include all the previous level characteristics: clear intention, focus, movement, and touch (see fig. 7.1 F). Think of Level Four as doing jumping jacks to make your point. Use grand and forceful gestures, and as mentioned, you may or may not include touch. When I use the words "grand" and "forceful," I do *not* mean you should strike a horse. Most often the message should be conveyed *without* touch. By following the previous steps in this book, you have already gained the trust and respect of your horse that make this kind of communication an extreme rather than a norm.

Level Four is exaggerated movement. You use big movements the same way a horse does to show the highest level of intensity when another is not responding to him at all. At times, you could find yourself in a dangerous place with a horse, and in these rare instances, Level Four intensity may be required to get a horse's attention, in order to be safe.

Conversations you've learned to this point are not meant to address extreme situations with panicked or angry horses. The Conversations you have should allow you to increase your level of intensity gradually until you get a response. When a human raises her intensity all the way up to a Four (let's say, striking the ground hard with the tail of a whip, making a cracking sound), you often cannot return to total calm (Inner and Outer Zero) in a matter of moments as is our goal, and this causes stress to the horse. The pressure from your body stays in the "on" position and he can feel it.

## The Difference Between Assertive and Aggressive

Humans are guilty of *thinking* about moments they aren't enjoying with their horses. Thoughts like; "Oh my gosh, this horse is never going to change!" "I can't believe it! I feel betrayed." "Why am I paying for this?" "He never understands me!" Those stories that you tell and retell yourself keep spinning the events round and round your head until you are no longer in the moment with the horse. This is an internal pressure that you experience, which horses are sensitive to.

Another form of pressure that you exude is when you think you are in a contest of wills and you must always win. On the contrary, I believe we must strive to *never* fight with a horse. Our predator instincts could unexpectedly trigger inappropriate action—that is, we might do the wrong thing, and it might harm our future abilities to be with and work with a particular horse. We also don't want to play the role of predator with a horse because he has great strength and accuracy in his own instinctual defense: kicking and biting.

In a Conversation, you do not want either your horse or yourself under the influence of adrenaline, one of the major stress hormones. Neither one of you will be thinking but instead simply reacting. Mother Nature will take over. When a human or horse has had a surge of adrenaline, it takes at least 20 minutes for that hormone to completely leave the body, unless he "runs it off."

### *Exercise:* "Assertive" Practice

To end the confusion between *assertive* and *aggressive*, try this exercise where you practice Level Four intensity *on the outside* while maintaining Zero *on the inside.*

**1** In an open space without horses nearby, hit a fitness ball with a whip using all the strength you can muster (figs. 7.2 A & B).

*7.2 A–C Use a fitness ball to explore Level Four intensity and allow yourself to emit great energy in one fluid motion (A & B). As soon as it's over, it's over (C).*

❷ Then go right back to Outer Zero (fig. 7.2 C).

❸ Observe your emotions as you do this. As you go from Zero up to Level One (forming intent), to Level Two (asking politely), to Level Three (telling, perhaps with touch), to Level Four (exaggerated movement), your emotional state *should not* change. But when hitting the ball in this exercise, many find it hard to perform the action and return immediately to Inner and Outer Zero. The trick is to allow your body the movement but keep your inner self "quiet"—like a samurai.

## Emotional Stability

To express large movements without emotion takes practice. But it is vital to be able to use large movements in Conversations, if needed, with *emotional stability*. Emotional stability is being able to identify and manage your emotions all the time you are with your horse.

It is not when you are waving sticks or flags, touching, or making large movements that progress happens with your horse. It is when the pressure *stops*. Keeping pressure on a horse is the fastest way to turn a *sensitive* Conversation into a *desensitizing* one. One horse would not grab onto another horse and simply hold on. To be the most effective, you need to be like a horse: work up the levels of intensity until you are heard, then immediately drop back to Zero.

*7.3 Jag and Luna buck playfully, showing "big" intensity, while Rocky stays Zero.*

If you observe any two or more horses turned out together, you will rarely see them use more than Level Two to communicate with each other. You will also see how they move from Zero to Two and then back to One or Zero almost immediately. Their goal is to spend life at Zero. Horses only rarely use Level Three or Four intensity with each other (fig. 7.3). By comparison, we may Groom a horse at a One or Two, and when doing groundwork with a horse, we may be at Three or Four intensity, which is telling and demanding.

## FINE-TUNING YOUR "X" AND "O" POSTURES

Let's think about some of the Horse Speak you've learned so far. Human-to-human body language is not all that different. When a stranger stares directly at you (Sending message—see p. 33), you may be uncomfortable. When a lover gazes into your eyes (Beckoning message—see p. 33), you melt. When someone stands  too close or directly in front of you, it feels different than when that person faces you slightly to one side. If someone poked your shoulder for no reason, you'd be confused or get angry. If a dance partner twirled you with a gentle hand on your arm, you'd turn back willingly into your partner's arms.

As we've discussed, the position of your torso, as well as how you hold or move your arms, legs, and head have great meaning to the horse. You give a powerful Sending message to a horse by forming degrees of "X" with your body.

There are many variations of the "X" Posture. A slight lifting of your chest is an "X," as is holding your head high. Any stance that exposes your torso is an "X." Lifting one arm to put pressure on a button on your horse is a Sending message and a form of "X." An extreme "X" is standing with your legs apart and arms held up and apart over your head, as if you are doing jumping jacks.

We all do some version of "X" instinctively with our horses. It is less instinctive to draw a horse to you with the "O" Posture. To remind you, in an "O" Posture, your arms form a circle low in front of your torso, you are naturally slightly bent over with unlocked knees, and your head is lowered. The "O" Posture is as irresistible to a horse as someone holding his arms out to you while offering a huge smile.

Some people tend to allow horses too "in" their Bubbles of Personal Space. Their body language is naturally "O-like": shoulders rounded, chest held in, knees soft, neck slightly bent. Other people can't catch horses because their baseline posture is too "X." Their chest is always out, shoulders always back.

With both these types of people, even though they aren't doing an "X" or "O" with their arms or doing it with conscious intention, horses are still sensitive to the

## Moving from "X" to "O" and Back Again
### IN PRACTICE

One of my students has an opinionated mare that, one winter day, got out of the pasture and was running around, quite animated and excited! In order to move the mare away from the barn where there was a danger of her being injured and to safely contain her in a paddock, my student had to drive her. She told me she used Level Three Sending messages, swinging a rope toward the mare's upper hip to express what she wanted.

Once the mare was in an "alley" between paddocks, her exit was blocked at one end. My student opened a gate to one of the paddocks, then positioned herself at the open end of the alley, holding a Zero, "How curious?" inner attitude to draw the mare toward her and the open paddock. She used a Level Four gesture, slapping the lead rope on the ground to turn the mare back away from her when it seemed she might try to blow past her, but she needed the mare to see the open gate that was nearby, so every time the mare turned back toward her, my student had a relaxed body stance and held the "O" Posture. If the mare kept coming straight at her with too much speed, she used a Level Four extreme "X" to turn her again. Alternately Beckoning and Sending eventually caused the mare to see the open gate and choose that option on her own. Horse Speak got the mare to start *thinking* and *stop* just reacting.

corresponding Beckoning and Sending messages and will respond. Now that we're playing with the "volume" of our intensity, it makes sense to use the intensity levels to fine-tune our "X" and "O" Postures.

## FINE-TUNING CORE ENERGY

Your posture influences how the center of your torso appears to the horse. In the "X" Posture, you stand up straight with a completely exposed belly button. In the "O" Posture, your belly button is drawn in and somewhat behind your lower arms. Remember that your belly button is a powerful communicator to a horse because behind it lies your Core Energy (see p. 68). This may be why the "X" and "O" Postures communicate so well.

You now have the tools to explore how sensitive your horse is to your Core Energy.

### *Conversation:* Managing Your Core Energy

❶ With the horse at liberty in his paddock or stall and you outside the enclosure, assume an "X" Posture of low intensity, and imagine light beaming from your belly button to a specific point on the horse's body. Count to 10. Your horse may respond immediately, or not at all. There is no wrong answer; you are just gathering data.

❷ Now, turn your body as if beaming an imaginary spotlight toward the direction your horse is looking. Look there as well, but keep your horse in your peripheral vision. Count to 10.

❸ Turn your Core Energy back toward the first point of focus on the horse. Observe if your horse shows any sort of reaction: a change in breath or ear position, a shift in stance or head height.

❹ Try this three times and don't forget to do Aw-Shucks, letting out a big breath, in between tries.

You can literally "open the path" the horse will move toward, or close him off to movement, by directing your belly button and your Core Energy, using the levels of intensity (figs. 7.4 A–C). When your horse is moving next to a wall or a fence, directing your eyes and Core Energy at the wall just in front of him can be like "closing a gate." You can then "open the gate" by swinging your Core Energy so it projects parallel to his own and his movement. Narrow his path by directing your Core Energy on a diagonal or at a 45-degree angle across the area in front of his chest. As we've already learned, horses can learn to back up when you turn your Core Energy to their chest. It is important to not only be aware of how

*7.4 A–C Using your "X" and "O" and Core Energy, and managing your intensity levels, can open the path you wish your horse to take (A) and show him where you want him to go (B). Lifting my arm higher, I can aim at Dakota's Go Away Face Button, telling her with an intense "X" to move that way.*

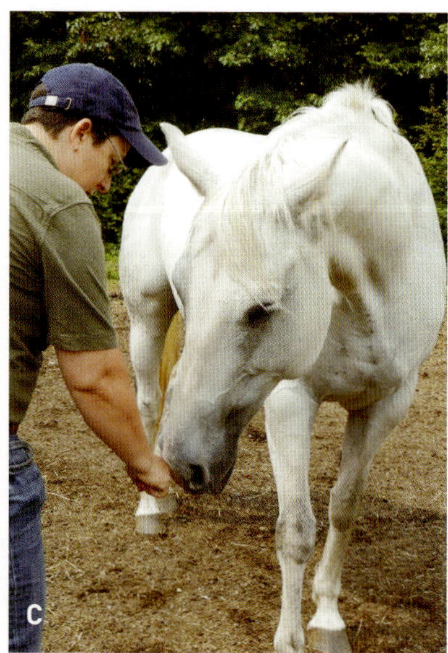



*7.5 A–C It is important to use your movements and Core Energy deliberately. I only ask Vati for one step back at a time (A). She comes forward to my Beckon (B & C).*

your belly button has a powerful influence, but to learn how to rate it using the levels of intensity (figs. 7.5 A–C).

Think about how we usually are with our horses. Most of us probably randomly direct our belly buttons and focus all over our horses' bodies, all the times we are with them! Since this all amounts to different forms of *pressure*, most horses will be confused by our communication. They know we're asking something of them but they have *no idea* what it is.

Your horse constantly moves his head when he is around you. Guess what? When your belly button is aimed toward his head, it is a Sending message. As soon as you turn your body to reach back and Groom the withers, for example, you remove the pressure and he will want to bring his head around toward you.

I've developed a Conversation to help you learn how your Core Energy influences your horse.

### Conversation: "Are You Ready?" "Yes, I'm Ready."

I've picked an arbitrary Button on the horse in order to practice this Conversation: The Go Away Face Button on the cheek (see p. 37). Of course, this Conversation can occur with *any* of the Buttons. The responsibility for this whole Conversation falls on *you*. It doesn't matter how the horse responds. You are simply watching for his reply, no matter what it is.

❶ Nod and step up to your horse holding out your knuckles to check in. Since you've been practicing Greeting for some time, your horse should immediately be engaged and curious.

❷ Now move back about 4 feet and stand at a 45-degree angle to one side of the horse so you can see one of his Go Away Face Buttons (fig. 7.6 A). You may want to scuff the footing to mark your starting place on the floor before you begin.

❸ Standing there, model your version of Outer Zero (see p. 9 to review). Your relaxed position might include a cocked hip or you may simply unlock your knees. Don't look directly at the horse but instead perhaps at the blanket hanging on the outside of the stall door or out the window in the stall wall behind the horse. Hold your arms loosely by your sides or rest your hands in your pockets. Think of your Zero *trigger image, song, or memory* for a moment (see p. 8).

❹ Next, with Zero emotion on the inside, stand up straight, moving into an "X" from your Outer Zero, and look at the cheek Go Away Face Button with Level One intensity. Your horse may lift his head, tighten his lips, or move his ears.

❺ No matter how slight his reaction, the *instant* he acknowledges you in any manner, exaggerate your return to Inner *and* Outer Zero by looking at the floor and blowing out a big breath. Play with this, and see just how *quiet* you can get.

❻ Now engage your Core Energy again and look at the Button on your horse's cheek. Again, *as soon* as he notices, look at the ground, rest one hip, and blow out a breath, finding Inner and Outer Zero.

❼ Look up, focus on your horse, notice his reaction, look down, and relax. At Zero, try to sneak a peek at your horse's face. You'll often get a much larger reaction when you *take the pressure off* than when you apply it.

In dialogue form, here's what this Conversation says between human and horse: "Are you ready?" "Yes I am ready." "Good, I'll go back to Zero." "Okay, I'll go back to Zero."

I often have horses lift their heads when I stand up, and then they relax as fast as I do. It almost becomes comical.

In this version of the Conversation, you move back and forth from Zero to Level One. You are working on calibration of expression when you look at a horse's Button. Level Two includes pointing at the Button, either with your own finger or

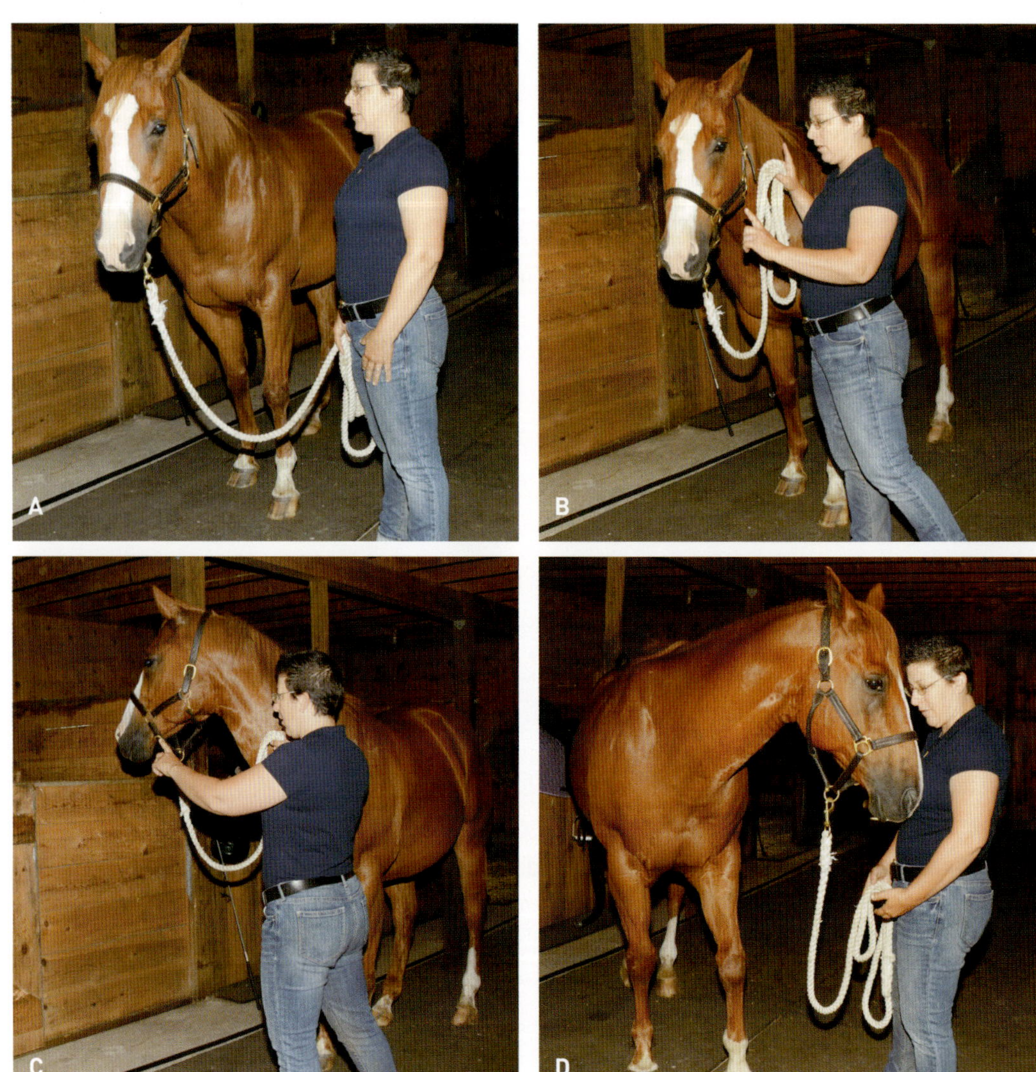

indirectly with a wave of a piece of string or a bit of cloth. Level Three includes characteristics of One and Two *with* an added step toward the horse and *maybe* touch. Four includes all the previous movements, now in exaggerated fashion, per- haps along with a touch with a finger or whip end, and another step closer so you can apply stronger pressure, maybe even include clucking noises.

To find clarity, going through all the levels of intensity at least three times. If your horse turns away, draw him back with a Greeting Knuckle Touch. I doubt he will because this should be really interesting to him! Clarifying your intensity levels dissolves a surprising amount of confusion.

Repeat the steps described in the Conversation on p. 111. Begin again at your

Inner and Outer Zero, engage your Core Energy, focus on the Go Away Face Button, and as soon as you see any reaction, drop back into Zero. But increase the pressure of your "X" so the dialogue is a Level Two.

> You: "May I ask you to move your head over?"
> Your horse: "I see you are asking."
> You: "Good, that's all I want. I'm going back to Zero."
> Your horse: "I'll go back to Zero, too."

Then *tell* him, then *insist* (figs. 7.6 A–D).

The point of this is not to move the horse's head or desensitize him to your request. The point is for you to learn how to be like a horse. This is *not* a Yielding Space Conversation. You are just practicing *phasing* your intensity levels. More importantly, you show your horse what you look like when you are phasing up and down the levels. Your challenge is to learn how to Retreat to Zero (see p. 32) faster and faster—the instant you see any flicker or tension from the horse.

## *Conversation:* **The Waltz**

This is a great activity when leaving the stall or returning to it.

**1** Stand slightly to one side of your horse's head, facing him.

**2** Focus on one of his front hooves with Level One intensity.

**3** With your foot on the same side, take a big step backward.

**4** Stop. No matter what his reaction, go to Zero and praise him.

**5** If your horse moved the hoof forward, focus Level One intensity at the *other* front hoof and take another big step backward. (If he didn't, simply try again.)

**6** Stop again and go to Zero.

**7** Repeat, saying, "One step," out loud, to coordinate your own micromovements. Have him come forward toward you, one hoof at a time, as you step back with the corresponding foot.

**8** Try suspending your foot in one step and see if your horse will hesitate with his own foot in the air.

 This is the beginning of dancing with your horse. Now look at the front hoof closest to you. Begin to lift your foot on the same side, this time move your foot *toward* the horse with Level One intensity. With practice, your horse will begin to "see" your intent and step back with just that one foot at the same time you move it forward. As you step that foot backward, returning it to its starting point, he may bring his foot forward. A bit like a waltz!

 Reward him with a breath and return to Zero, even if he just looks at your foot—he's trying to figure out what it is you are asking.

*7.7 Adding a bow to Fun with Feet— just for kicks!*

Some horses are more sensitive than others. When trying this Conversation with a less sensitive one, you may need to indicate the hoof you are focusing on with a crop. Hold the crop in the hand on the same side as the foot you are planning to step with. Start a little farther away from the horse, still slightly to one side. As you pick up the foot you are going to step with, point the crop at the front hoof you want him to move forward or backward. If you have to go to a higher intensity level, just tap on the hoof wall with the end of the crop.

When your horse moves the foot, take a big breath and go to Zero before you engage your Core Energy and deliberately return to your starting point with the foot that you moved. You can begin the dance by stepping back with your own foot to bring the horse's foot forward, or by stepping forward and asking the horse to step back. Either way is fun.

Once your horse understands and begins to enjoy this as much as you do, practice it standing at his shoulder, facing the same direction he is. Move your left foot to cue his left, and your right to cue his right hoof. You can step back and forth, or back several steps, or any combination, and the horse will follow your feet with his. Eventually, you will be able to cross-step next to your horse, and he will step one leg across his other, moving sideways along with you.

You will do more with these Fun with Feet Conversations later at liberty (see p. 162). When you play with your horses, it teaches you to be more lighthearted. Bringing this happier attitude to your riding and work with horses will benefit you both (fig. 7.7).

## PHASING DOWN

Being at Outer and Inner Zero (physical and mental) in front of a horse is a practice in mindfulness. With what you've learned in this chapter you'll begin to notice when you have phased up in intensity but didn't really mean to ask anything—and you can catch yourself and Retreat to Zero.

As well as I understand horses, I still am human and know how we think. Many of us will take away from this discussion how it feels to *phase up* our intensity, and how to correlate our movements to each level. If, on the other hand, we think like horses, we'll be more interested in the *downward* phasing of intensity. Remember, horses look for any excuse to "phase down" and have peace and quiet.

How many of us have been taught that all the time we are with our horses, we must stand tall and stay focused to show our dominant position? In the horse's world, it is just the opposite. The Alpha horse is powerful enough to move a number of horses at once with just a glance. However, the Alpha does not keep the pressure on when a request has been answered. The leader immediately Retreats to Zero.

Keeping pressure on a horse, even inadvertently, is the fastest way to create resistance. When you *release* the pressure you gain respect. To create a delightful connection, you must gain self-awareness. Any resistance you meet tells you how the horse feels about his relationship with you. If he moves too fast in answer to a request, he is not really listening, he is just trying to avoid you or rush an answer for fear of getting it wrong. If he moves too slowly, he is stoic, "sticky," or stubborn. With stoic and stubborn, you have to reward more often until he "gets" it. No matter which end of the spectrum your horse is on, you must release pressure and *phase down,* showing the horse you are as effective, fair, and considerate as any horse would be.

When your body has memorized how to phase up and, more importantly, how to *phase down* in intensity, you will become more subtle. It may not look as if you are accomplishing much in the Conversations in this Step, but after having them several times, your horse will be able to "see" your Zero. He'll know how you look when you are *not* putting pressure on him. He will know what your body looks like when you are asking a question. He can relax around you and feel safe because he knows how to read you and how to predict your movements.

## FINE-TUNING APPROACH AND RETREAT

In Step 1 we first discussed Approach and Retreat and its uses in Horse Speak (see p. 32). Now it is time to use what we know about intensity levels to improve the way we use this technique.

### Frightening Object

Horses naturally Approach and Retreat from frightening objects. This means we can apply this Conversation to change a horse's mind about something he's unsure of. Say your horse does not like fly spray. At Zero intensity, inside and out, hold the fly spray close to the horse. When he reacts, quickly take it away. The fly spray is the object of pressure but *you* are *not.* You must not add extra intensity from your own body to the pressure of the fly spray. Now that you've been through the previous exercises, your horse knows what you look like when your body language is at Zero.

Now your horse will be more willing to participate in stages of pressure presented by the bottle of fly spray, so you repeat Approach and Retreat with the bottle

*7.8 A–C  I ask Vati, "May I approach you with this fly spray?" Because she is looking at me, I get a "Yes" (A). I retreat to say, "Thank you" (B), and repeat Approach and Retreat until eventually I can spray her (C).*

nearby, touching him, with the spray pointed away from him into the air, and eventually, on him (figs. 7.8 A–C). Always, the important thing is to take the pressure away as soon as the horse reacts. This builds trust—he see that you are sensitive to his replies. Though you may have to repeat this Conversation, eventually he will not be concerned at all about the bottle of fly spray!

## Injury

You can use an Approach and Retreat pattern with an injured and frightened horse. It might look like this: You glance at the wound and take a breath, then look away from it. Look at the ground before you look back. Move into the horse's personal space and stroke his body above or below the wound. Look away again (that is, Retreat). Breathe again, gently, gradually, moving closer to the wound, repeating this Approach and Retreat Conversation as you clean, treat, and wrap the wound.

## Catching

Approach and Retreat is also effective way to catch a horse. Approach, and the moment he looks at you, go to Zero, look at the ground, and do Aw-Shucks—maybe even say out loud, "I don't want you." This Retreat on your part caters to their curious nature. I've had runaway horses come to me when I've done this. They were fascinated by my actions! Knowing how to Retreat to Zero and regulate your body intensity can be the keys to changing the ways of "hard to catch" horses.

## FINE-TUNING GO AWAY AND COME BACK

Remember how we learned the difference between Approach and Retreat and Go Away and Come Back in Step 1? Approach and Retreat is the way you enter another's space, and Go Away and Come Back is how you send someone out of your space or ask for her to return. I have developed Conversations similar to Approach and Retreat, except now you are asking your horse to back up (Go Away) then come to you (Come Back). Not only is this probably the first Conversation he ever had with his mom, but all horses understand being pushed away and invited back as it is what they do to each other all day in the cohesive group that is their herd.

The idea is for you to see how small a Sending message it takes for you to have the horse move away from you—even the slightest lean or shift. Then, see how small a Beckoning message you can make to have him return to you. You've practiced managing your intensity levels so you're ready to see how subtle you can be.

### *Conversation:* This Hay Is for You

Another great Conversation can happen over a pile of hay. Using hay as a Conversation tool is very effective, but be mindful that some horses have issues around food, so safety first.

❶ In a paddock, scatter a few leaves of hay and stand on them.

❷ Beckon your horse to come to you from wherever he is—presumably near the hay (figs. 7.9 A & B).

❸ As soon as he even *looks* at you, Retreat to Zero, leave his space, and let him eat (fig. 7.9 C).

❹ Return to the hay in "X" Posture, looking to now move him off. However, look for a small indication of his willingness to Go Away—such as an ear flick—rather than driving him away completely.

❺ Repeat this, demonstrating Inner and Outer Zero as dramatically as you can: Pretend to eat hay, breathe slowly and deeply.

Fine-tuning go Away and Come Back gives you a good idea about how your horse will respond to your need for space or invitation to approach. Some will return to you with tact and gentleness. Some will come too fast or crowd you. Your

*7.9 A–C I use my "X" Posture to politely move both horses off the hay pile (A). You can clearly see the natural arcs and circles horses make (B). I Beckon to them, and they both check in with me before eating again (C).*

horse may not know what intensity level he should be in around you. This is part of what we are working on here. By using Horse Speak in a Conversation rooted in horse protocol, most horses will be open to a Send or Beckon, even if only a few steps. Take what is offered.

### *Conversation:* Go Back, Come Forward

**1** Stand in front of your horse, just to one side of his head, facing his tail.

**2** Ask him to back up, using Level One intensity, and focus on the Back-Up Button (see p. 39).

**3** Phase your intensity upward, pointing, gesturing toward, or even touching the Back-Up Button if needed until your horse shows *any* reaction or a willingness to yield.

**4** Even when the horse's response is simply to lean backward, go back to Zero. You must always acknowledge a "try."

**5** Now step backward and change into a Beckoning Come Back pose: bow slightly and form your "O" Posture, making a circle with your arms. Bend your knees into a slight crouch. Say: "Come here!" in a lighthearted way.

**6** As your horse leans or comes toward you, drop to Zero on the outside and in. Reward him with an Aw-Shucks or a big-breath sigh.

**7** Repeat this three times and observe the difference in his response once the horse knows what you are asking.

When you are comfortable with the steps, try this Conversation inside your horse's stall with him loose or in a halter and lead. Holding Conversations with him in his stall may create less resistance when you take him other places: the stall, aisle, driveway, and arena present different *Thresholds* (see p. 62). Each location has preconceived associations for you both.

### *Conversation:* Beckoning for Bonding

**1** With your horse is at liberty, explore the Follow Me Button. Press lightly but firmly with your whole palm on his upper neck for a second or two (see p. 37).

❷ Then turn and walk slowly away. In horse language, you've just asked the horse to follow you. Your horse may be amazed you know about this and what it means. His mother used it, and it encourages trust and connection. If he doesn't follow, return and try again. Chances are, the horse will at least follow the pressure of your hand as you remove it, moving his neck toward you.

When you see your horse holding his head low and with sideways ears, he may be trying to Beckon another horse to him. This is the horse equivalent to your "O." It is also the gesture of apology he makes to another horse. Let's say one horse is really crabby and drives another horse away from the hay pile with more force than necessary. You may see the offending horse apologize with sideways ears with a low head and soft face, eventually transitioning to bringing his ears more forward and reaching his neck long and low, he is making the horse version of "O."

This gesture is also how your horse draws you to him. When he does it along with a visible "huffing" with his nostrils, you may start to feel a fluttering in your core, or a tingling sensation in your heart or stomach. Your autonomic nervous system is responding physically to his Beckoning. Whenever you are with any animal, you are influenced by this phenomenon.

Some horses will follow you right away when asked. Others want to follow but are afraid—they might have been punished for following a Beckoning cue in the past when a human did not realize what she was "saying" in the language of the horse. You may notice your horse is "stuck" with a lowered head, relaxed ears, and a yearning look on his face, as if begging for a treat—yet he still cannot bring himself to move. Just as you've tuned into your own subtle levels of intensity, so you should tune into his: His intention to follow you might be indicated with the slightest lean. Praise your horse for the effort and praise yourself for noticing! Caution: When a horse follows you, he may think you are inviting him to drive you forward, and he may flatten his ears. Keep an eye on facial expressions if your horse is typically Alpha. Be prepared to turn and do "X."

### *Conversation:* **Horse Hug**

Another Beckoning Conversation can turn into a hug. One horse wraps his neck around another to express deep friendship.

❶ Start exploring the Horse Hug by standing with your back to the Shoulder Button on one side (see p. 38). Your body position tells the horse you trust him.

**7.10** *To begin a Horse Hug, reach one hand under the horse's jaw and rest it lightly on his opposite cheek. This can make him nervous, so keep yourself at Zero.*

❷ Reach one hand to the horse's cheek on his other side (fig. 7.10).

❸ With your free hand, scratch under his jaw bone, tickle his chin, or softly stroke under his throatlatch.

❹ As you do this, encourage the horse to bring his head toward you and across your body with the hand that is on his cheek. He may well pull his face away because this is similar to how a predator would flip the horse from under his neck. Keep a, "How curious?" attitude and notice if he braces against your request. Don't put a lot of pressure on him but also don't quit easily.

❺ Ask at least three times on each side. Don't be in a hurry. Just hang out with him, and when he does yield his face in your direction, even a little bit, release immediately and praise him.

❻ When he wraps you in a full hug, reach up with your free hand and stroke his forehead at the Friendly Button at the base of the forelock (see p. 37). You are not training him to do a trick or forcing anything on him. You are just encouraging the horse to hear your request. I've had horses Yawn into my hands during this Conversation!

❼ Sometimes, you'll see one horse take a big sniff of another's neck. This is a Beckoning Conversation on an emotional level and a step you can add here. When you sniff your horse's neck, you tell him you *really* like him. Your horse may relax his ears sideways when you breathe in his smell at the neck. When he allows you to hug his neck and breathe in his scent, your friendship will deepen. I sometimes have my students sniff their horses' necks before they ride.

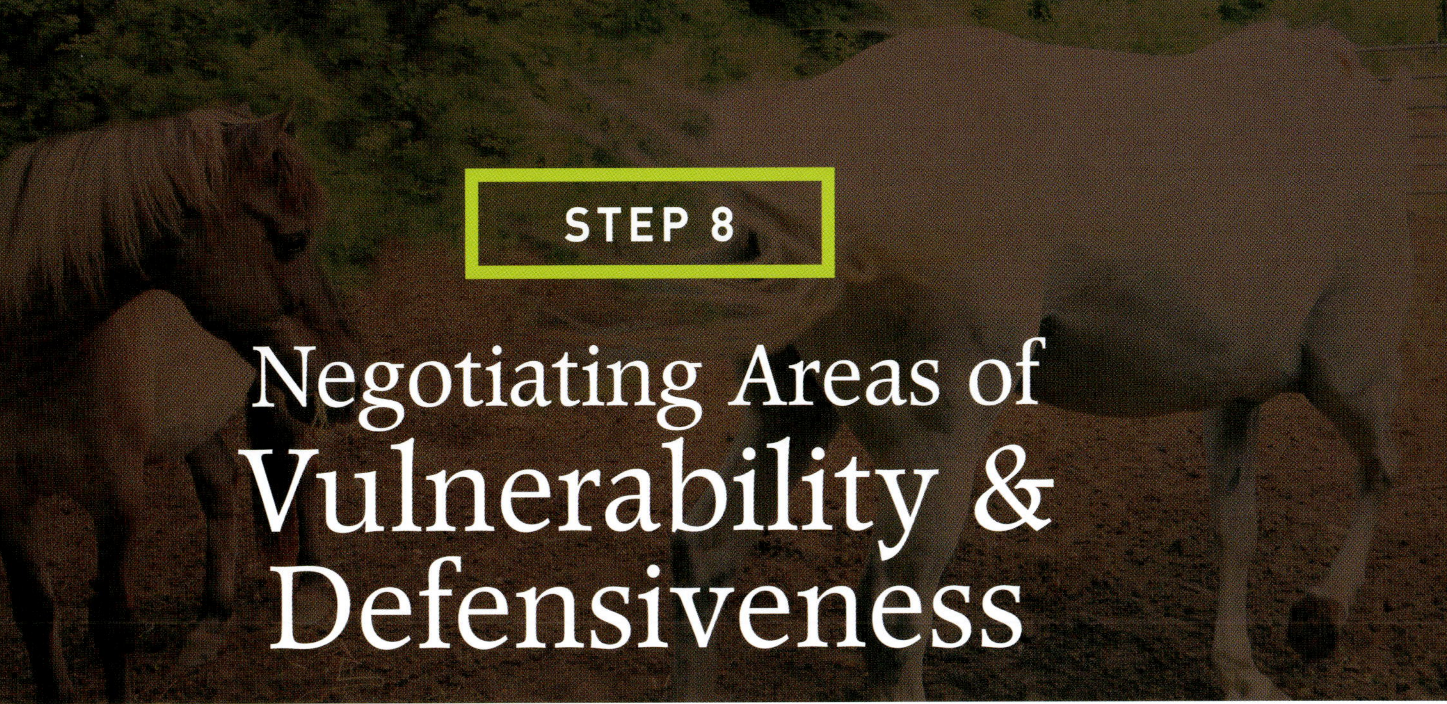


# Negotiating Areas of Vulnerability & Defensiveness

## GETTING TO KNOW THE HORSE'S VULNERABLE BITS

The most vulnerable parts of the horse's body are the back and belly. They are suspended between the striking and biting ability of the forehand and the powerful defensive kick of the hind legs.

The horse expresses friendship and *unity* by allowing other horses access to these unguarded areas of his body. Watch for horse friends that stand alongside each other, sharing quiet time and mutual comfort. We've all seen horses walking parallel to each other, sometimes with Matching Strides. Horses rub their heads against other horses' bellies. The head of a foal is often under his mother's belly to nurse, and the stallion smells another's underbelly. All are expressions of the intimacy horses can extend to each other.

We have a compelling relationship with this part of the horse's body as we strive for the ultimate unity when mounted. If a horse does not willingly let you have access to his back and belly, he does not feel safe with you. Most domestic horses have been taught to accept us being in this area to brush, girth, and mount. However, when a horse has a "hard" look on his face or is holding his breath during our interactions with his back and belly, he is conveying distress and discomfort. Especially when trapped on the cross-ties, the anxious or uncomfortable

## Keys to Horse Speak: Step 8

Back and Belly (p. 123)
Girth Button (p. 124)
Jump-Up Button (p. 125)
Making Friends Behind (p. 125)
Punctuation (p. 127)
Gone (p. 129)
Tail Swish (p. 129)
Hip-Drive Button (p. 131)
Turning the Canoe (p. 133)
Yield-Over Button (p. 134)

horse may cowkick, sidestep, fidget, or look "tucked-up," trying to protect this most exposed and vulnerable part of his body.

It only takes small consistent efforts to "make friends" with the belly of a horse. If your horse is defensive about a place on his back or belly, begin a Conversation about it in a stall with a halter and lead rope. Touch it softly, turn away, then touch again three times in an Approach and Retreat Greeting. Begin with gentle scratching with finger pads or brushing his belly with a soft face brush. Watch for any message of discomfort. A swish of the tail, stomp of the foot, or shift of weight shows you it is time to Retreat to the last area of contact where he was comfortable, and to use less intensity when you Approach again. Touching the underbelly is not a "test" to see if your horse will accept you or not. It is a Conversation to first gain and then express *unity* with this sensitive area.

## The Girth Button

There is a sensitive area located low and a bit behind the horse's girth line (fig. 8.1). One horse can move another forward, backward, or sideways just by pointing his muzzle at this spot. This is the Girth Button, and it is a powerful means of claiming space. In horse-to-horse communication, it can also be the "last resort" Sending Button in order to claim sideways space.

Horses are constantly negotiating parallel space between them, drawing each other closer or creating distance between each other's girth lines. They are sensitive to this space, and when they are not friends with one another, they won't line up their bodies to "match" at the girth line. We have all experienced this when trying to ride side by side with our friends on the trail. When horses are running in a herd together, I believe their girth lines contribute awareness that helps meter the distance between them. Horses steer each other with subtle gestures of the head and neck as well as with expressions of the tail and rump. The results of these gestures are expressed in the distance between them at the girth line.

The Girth Line Button is a horse's "Balance Button" and also the pivot

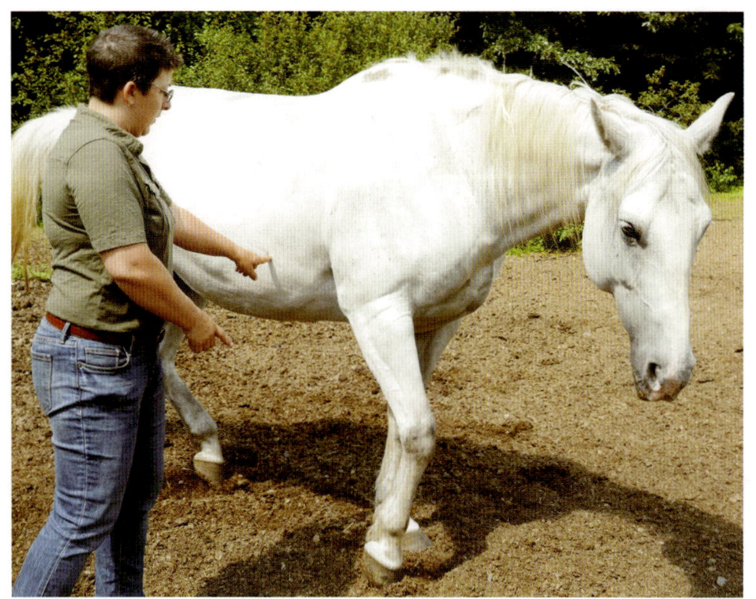

*8.1 The Girth Button asks for sideways or forward movement.*

point for the whole body. It largely corresponds to the location of the diaphragm of the horse. It is possible, then, that to horses, girthing feels like a chokehold that restricts their breath. There are many techniques that help horses feel better about the tightening of the girth. My favorite method is to use my breath.

### *Conversation:* **Breathe Up the Girth**

**❶** After saddling and loosely attaching the girth, stand next to your horse at Inner and Outer Zero.

**❷** Take several large in-breaths, releasing audible out-breaths. You are taking the time to ask your horse to relax as much as possible.

**❸** Keep breathing deeply as you gradually tighten the girth. Never take for granted the permission a horse extends by allowing it.

## The Jump-Up Button

Moving back from the girth line, along the belly under the ribs on each side of the barrel, are Buttons that tell the horse to "jump up." When a predator attacks from the ground, he might try to catch the end of the horse's rib cage in his jaws. Horses are very sensitive to any stimulation in this area, even the touch of a brush or a stick might make them leap up or forward to "get away." When you ride, your heels can contact the Jump-Up Button. Spurs mimic the sharp touch of a predator, so they stimulate the "go" reflex. I never wear spurs because this area of the horse's body is so especially sensitive.

*8.2 Further down the barrel from the Girth Button is the Jump-Up Button, which tells the horse to scoot over or move forward faster.*

## MAKING FRIENDS WITH THE DEFENSIVE END

Anytime I approach the horse's hindquarters, I Greet this part of the horse with three touches—usually in the area of the gaskin. Then I might Rock the Baby (see p. 55) on the rump or dock of the tail, ask for some small yields or leans away from pressure using the Yield-Over Button just above the stifle (we'll talk more about this shortly—see p. 134). I keep my peripheral vision attentive to my horse's facial and tail messages. I also watch for how he is breathing. I "make friends" with the hind end *before* I try to handle the hind feet.

### *Conversation:* "Making Friends" Behind

Just as you practiced a slow Approach and Retreat in the very beginning of this book with the horse's face, now you will do the same with his hind end.

**❶** As you approach the hindquarters, watch for a twitch in the tail or a shift in your horse's weight.

## "Do No Harm"
### IN PRACTICE

It is within the parameters of their social structure for horses to go to Level Four intensity and deliver a kick to another horse when necessary. Isn't it interesting that on any given day a horse can "lay into" another horse with full intensity, yet even under great distress, most horses never threaten us with the same level of assertiveness? Even when I was bitten or kicked in the past, it was with nowhere like the power I see horses use on one another as a matter of course. This goes to show that when a horse *does* attack a person with actual intention to harm, something has gone terribly wrong.

Why, exactly, do they *not* harm us on a daily basis when they have every ability to? How is it that even after being mistreated, abused, or neglected—like so many of the rescue horses I have worked with—they consistently retreat and first attempt to avoid us, only trying to drive us away as a last resort? I have come to believe that horses have in their social agreement with us a "Do no harm" clause. They must, or we couldn't have survived with them for thousands of years.

Some people feel it is okay for us to strike horses, first because they strike each other, and second because we can never hit one as hard as another horse would kick him. But as I've tried to make clear in this book, horses express all kinds of subtle language *before* they kick or bite one another. What leads up to the kick is their Conversation: a tail flick to show annoyance; a shift of weight that looks like a slight lean to free up a leg; or even the lifting of a foot as a more direct threat. These are all early-warning cues *before the strike, which comes with surgical precision.* The point is, there is no need to resort to striking a horse, because learning Horse Speak keeps you calm and teaches you how to "hear" what the horse is saying, which in the end will make you safer.

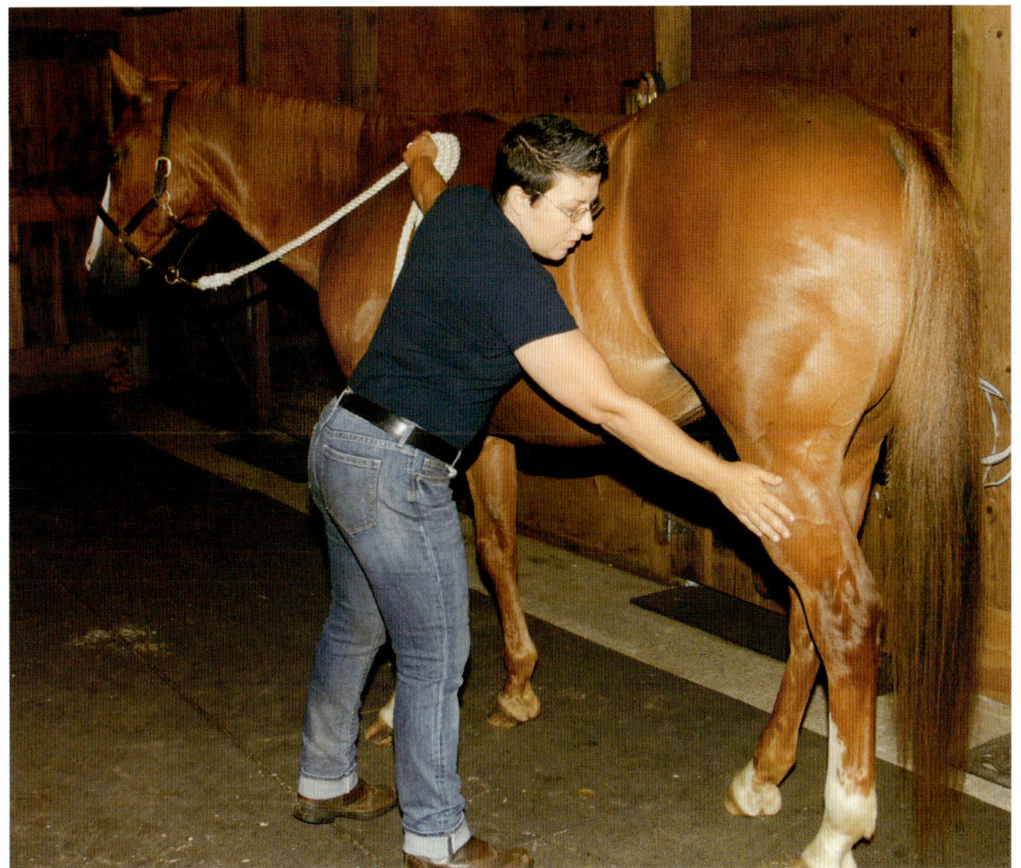

*8.3* *I have one hand on the withers for the horse's comfort while the other massages the gaskin, prior to my handling the hind feet.*

❷ At the edge of his Bubble of Personal Space, stop, do Aw-Shucks, nod, and then continue your approach. Each time the horse reacts, Pause to tell him you see his Bubble and respect it.

❸ Start Greeting with a touch at his gaskin area, then stroke down the leg to the hoof three times (fig. 8.3). Do Aw-Shucks or blow a long breath out between these touches. The gaskin is the bulging muscle above the hock on the outside of the hind leg; it has important pressure points that aid digestion and the stomach. Sometime, just gently scratch your horse's gaskin and see what kind of response you get. I believe I've prevented mild colic from worsening with vigorous massage of the gaskin.

## Punctuation

When venturing anywhere near the hindquarters, even in the initial Greeting as described, beginning on p. 126, you must consider the horse's tail. Tail gestures and hindquarter positions are usually just punctuating a message that began with the face. So, when you don't notice your horse's facial expressions, you are given another chance to understand his opinion—perhaps more clearly conveyed with his hindquarters and tail. When I know a horse has a tendency to kick, I watch his face carefully. In fact, the hind end and head are connected in more ways than one. One end of the horse can actually show symptoms of issues at the other. For example, if a horse is fussy when I handle his head, I might explore the possibility of pain in his hindquarters. Healing modalities directed to the "opposite end" of your horse may offer an alternative access to the root of a problem.

Once you start to see Horse Speak as it is expressed in a horse's tail you'll be flabbergasted you did not see it before. Be aware: a horse that has minimal control of his tail, or a limp, inexpressive tail, may have a spine injury.

- Lifted, slightly arched tails are an expression of confidence—you often see this depicted on equine statues (fig. 8.4).

- We've all seen a horse "flagging" his tail, holding it upward so the hairs fan out like a flag in the wind (fig. 8.5). This communicates either fearful or joyful excitement. I've also seen young mares flag their tails when being chased by other horses. It is a gesture that almost says, "Don't hurt me, I'm just a girl."

- A tight or clamped tail communicates fear, tension, even anger. A horse tucks his tail much like a dog. He clamps his tail when he feels threatened, has pain or tightness in his rump muscles, or is just about to kick out (fig. 8.6).

- A horse can "wring" his tail when in pain or distress: The tail rubs the body and moves back and forth in small windshield-wiper motions, or in a circular motion.

*8.4 A lifted "confident" tail.*

*8.5 A "flagging" excited tail.*

This can give you the first indication of colic or founder before any other symptoms present.

- Horses "wag" their tails when happy and content. You see this when two horses are standing near each other, just beginning to nap. Though there might not be any flies, their tails will still have a relaxed, side-to-side swing.

- A horse can do a single, smallish swish at a time when he is concentrating, frustrated, or annoyed (fig. 8.7).

- I've seen performance horses do a huge tail lift and downward, sharp swish after jumping a fence, often snorting an exhale along with this movement. I always think of this as an exclamation point: This tail is saying, "Yes! I nailed that one!"

- Horses seem to express rhythm with their tails in performance. It is like they are keeping beat with their movement, much like we clap or snap our fingers to music.

*8.6 When the tail is clamped, the horse is very defensive.*

## Gone

With the tail we come to the Fourth G of Horse Speak: Gone. In fact, perhaps the most important tail position that is easy for us to see is the, "I'm Gone," or "I'm done," message. This huge, strong swish is an accentuated version of the

*8.7 Mama (on the left) swishes her tail once as she cuts through Jag's Bubble of Personal Space. Jag swishes back in response.*

**8.8** *Rocky puts pressure on Vati by looking at her. Vati gives a sharp flick of her tail in protest and is Gone.*

concentrated-frustrated single swish directed toward another horse, or you, that says, "I'm so done with this and with you" (fig. 8.8). As we've discussed, it is the horse's period at the end of a sentence. It is also his "No." Be mindful of the horse's "No." Knowing when it is his wish to be Gone or to leave a Conversation enables you to use softer Horse Speak messages; when to check your intensity; when to go to Zero; when to take your Core Energy pressure off; and when to Scan

## Gone
### IN PRACTICE

One cold, rainy day, a student of mine went into her mare's stall to put her horse's "raincoat" on. The weather had been great for months with little need for blanketing, and my student saw her mare say, "No," with her tail as the horse turned away. My student Paused, leaned against the front corner of the stall and looked out to the aisle, extending one hand knuckles up in Greeting, and did some deep breathing. She calmly said out loud, "You can't go out on grass without your jacket today," and in a moment, she felt her mare's muzzle on her knuckles. The two completed the Greeting Ritual (see p. 52). Then her mare quietly stood still as she was blanketed.

Just acknowledging the horse's "No" can change everything. My student did not get angry or use force to blanket her mare. She had all the conversational tools she needed to understand her mare's feelings and talk back to her about the issue.

the Horizon (for example). Gone is a cue for you to look at what he could be "hearing" from *his* perspective. For us to have real "horse to horse" Conversations, we need a way to say Gone, too.

## *Conversation:* **Tail Swish**

You can mirror the Gone Tail Swish gesture.

**❶** Swing your arm and hand across your backside or thigh, once, briskly, as if shooing off a fly. At times, you will need to use this to let your horse know you are finished with a Conversation. It is how you can say you are done with something and it is time to move on.

## Hindquarter Sending Messages

When horses need to shift location in the herd, it is often accomplished with a series of subtle Sending messages from one or some horses, and by other horses following each other. The "drivers" of the herd tend to be higher in the pecking order. They are also automatically responsible for the safety of the horses being driven. In documentaries about and live footage of wild horses, we often clearly see a band stallion lagging behind his herd to drive a weaker member—for instance, a young colt or an older mare—forward. He is in the position to not only keep his herd together but also to fend off predators. Any horse that is Sending says to the horse being driven, "I am responsible for your safety, I've got your back." Even if there are only two horses in a herd, this still holds true.

Horses have deep basic social needs to both follow and to be driven by others from behind. Much of our interaction with horses taps into these instincts. Some skilled trainers intuitively pick up on the horse's need to be driven. When we Send a horse forward with pressure toward the hindquarters, as in certain forms of round-penning and longeing, for example, we should be aware that Sending is linked to protection and responsibility—that is, making the horse feel safe.

## The Hip-Drive Button

The Hip-Drive Button on the horse is one of the most sensitive Buttons: horses ask each other to move forward or sideways with Sending pressure toward the hindquarters all the time. These messages are directed to the Button located at the apex of the hip. The Hip-Drive Button is also a psychological responsive button linked to herd cohesiveness and safety. Many horses will move off just because another horse is looking at this Button; horses are hard-wired to respond to pressure in this area. Domestic horses do not have to worry about predators, nor do they have to travel distance in order to find food, but the essential leading and following Conversations are still present in petty arguments over hay, water, or the best shady spot. As you develop your connection to it, you, too, will be able to influence this Button with just a look. It is a powerful way to move a horse.

This is also the most misused Button when it comes to human cueing. Humans can push horses around with it in the same way "horse bullies" do. When a human bullies a horse by overusing this Button, it is almost predatory, and the horse often goes into a state of inner submission, glazes over, and shuts down. Although he may look "trained," on the outside, this horse is "caved-in" and performing in a robotic, inauthentic way. In a crisis, this horse will strive to keep himself safe, rather than looking to the human who is driving him.

Today, people are beginning to sense they shouldn't be predatory bullies, abusing the Hip-Drive Button on their horses. But many still don't know how to serve the needs of the horse that wants to feel *protected* by being driven. We can create soft, yet effective Conversations that are fair and thoughtful with this Button. We can Send our horses without aggression and not give into our predatory instincts to chase our horses. Instead, we can say, "I've got your back and you can relax now." This is the difference between *true acceptance* and temporary submission.

Many of us don't know any other way of driving our horses other than chasing them. I was guilty of this when I first learned to use the round pen. But once I realized how to have a Conversation with the Hip-Drive Button and *not* chase my horse,

*8.9 A–C At Liberty, I begin at Vati's Shoulder Button and work my way toward the Hip-Drive Button (A). All I need to do is point one finger at the Yield-Over Button located near Vati's stifle and another at her Hip-Drive Button to get movement (B). As always, I use an "O" and go to Zero for the Come Back so we can check in (C).*

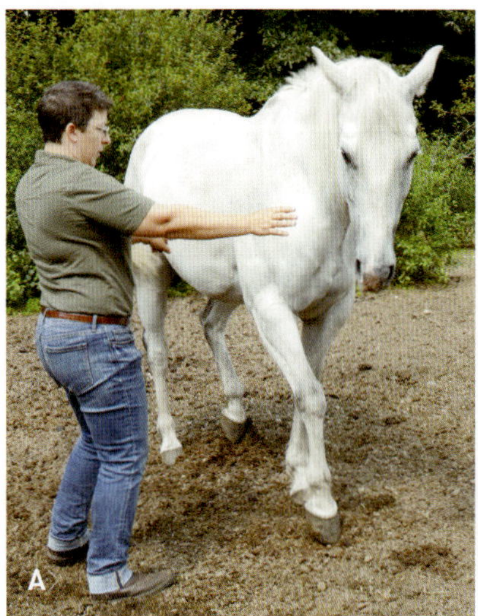
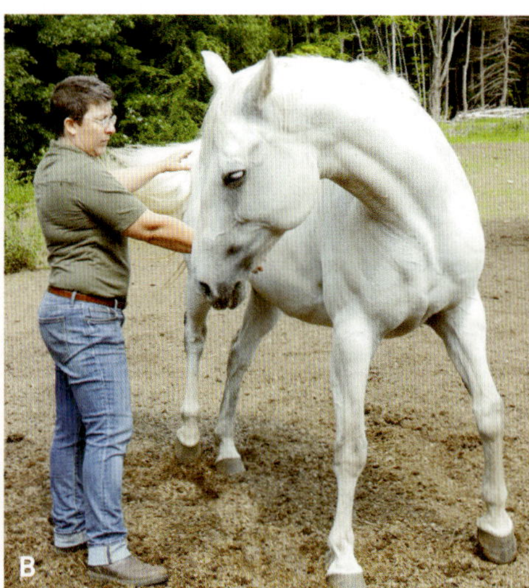
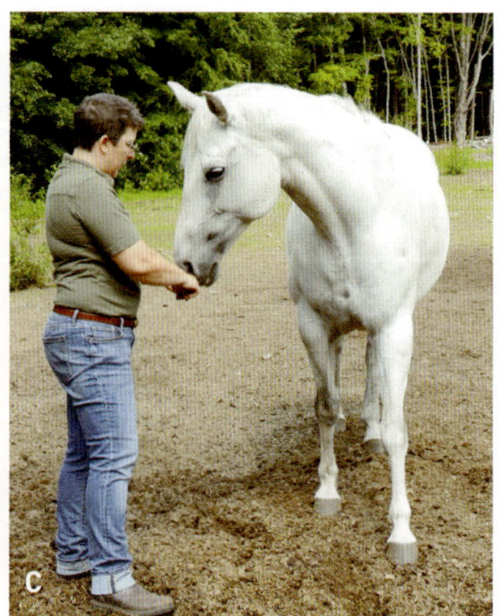

I felt a sense of shame. I'd been so wrong before. With my new understanding of this Button, I feel the rewards: an open attitude from my horses and a more thoughtful way to create *acceptance*. I have found a better path in Horse Speak for all Hip-Drive Button Conversations, whether on the lead or longe line, from a distance, in the stall, or at liberty. We can have Conversations without aggression, without bullying, and without chasing.

## Conversation: Send

**1** In the stall, paddock, or field, use your Core Energy to "hold" your horse as you move into position near the Hip-Drive Button at the top of the horse's rump (fig. 8.9 A).

**2** Direct your Core Energy (point your finger, if necessary) at the Hip-Drive Button to ask the horse to move (fig. 8.9 B).

**3** Finish the Conversation with an invitation to Come Back to your "O" Posture (fig. 8.9 C).

You can also turn a horse around to face you with pressure on the Hip-Drive Button. It is like turning a canoe in water: you push on the back end of the canoe and the front swings around to face you. The mare teaches this to her baby: when the foal is facing her, she has his attention, and she may now walk off, using her own flank as a target for the foal to follow.

## Conversation: Turning the Canoe

**1** With your horse facing away from you, aim your Core Energy at the Hip-Drive Button (fig. 8.10 A).

**2** Then use your "O" Posture to Beckon the horse's front end toward you (fig. 8.10 B).

**3** End with a Knuckle Touch and Zero (fig. 8.10 C).

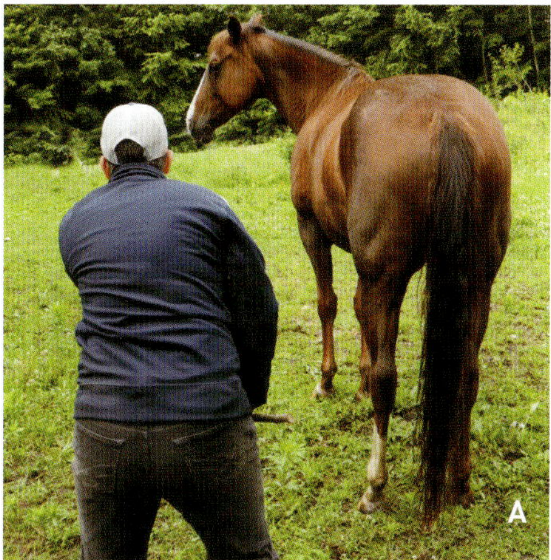

*8.10 A–C  Jag is facing away, and I want her to turn back toward me. I aim my belly button at her Hip-Drive Button (A). As my "O" Beckons Jag to me, you can see her pivot point in the center of her body (B). This is what I call Turning the Canoe. When I stand up straight but hold Inner Zero, it tells Jag to stop. I hold out my Knuckles to Greet her (C).*

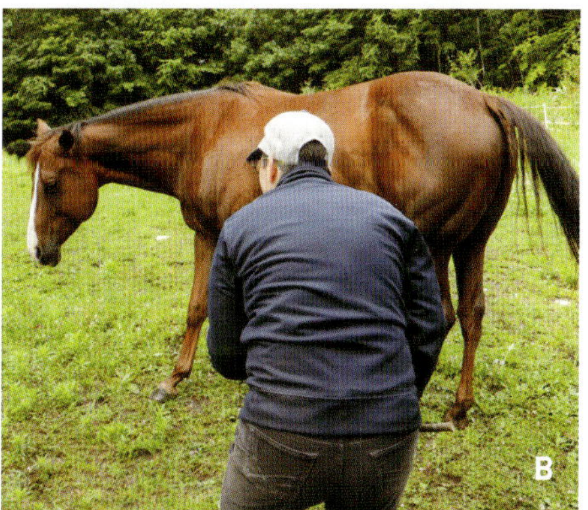

With the Hip-Drive Button, the request is, "Move over and take your feet with you." No matter what intensity is used, the message is the same: Asking the hind-quarters to step over says you want your Bubble of Personal Space respected in regards to the hind end's range and reach. The more you ask any Button to yield to you, the higher ranking you gain within the "herd."

## The Yield-Over Button

Whether asking the horse to pivot or driving him forward, Sending with the Hip-Drive Button furthers your relationship and understanding of your horse. There is also a wonderful way to rebalance the horse mentally, emotionally, and physically via a slight sideways yield. As we've already touched upon, sometimes, horses can get "stuck" or confused. In these cases, I ask for sideways movement using the Yield-Over Button in the hollow at the top of the stifle.

### *Conversation:* Step Over and Rebalance

❶ Ask for the yield with your Core Energy just until you have a lean or a step-under of the hind feet, causing the horse to have to rebalance his whole body (fig. 8.11).

❷ Then return to Zero and take the pressure off.

This little Conversation on the ground translates into yield of the haunches when you are mounted.

*8.11  I use just enough intensity to ask for a yield of Rocky's hind end— notice his right hind leg stepping under, his low head, and his calm expression.*

# IN HAND

JOE'S PRIOR TRAINING before he met me targeted his *physical* movement. The Conversations I've had so far with him told me what he does not understand and where he reacts emotionally. I return to work with Joe's leading skills and some of his basic ground manners. Mike and Liz, his owners, say they feel better about their interactions with Joe, and he seems genuinely interested in everything they do with him. They have a new understanding about his needs, wants, and intentions.

As I approach Joe's stall, I can see how his facial expressions compare to my last visit. His eyes are brighter; his head, while still held high, is facing me as soon as I enter the barn. He has already started breathing in my scent. I begin breathing in his scent by huffing my breath in his direction. I do three huffs, then I snort/blow through my nose. I make the Nurturing Breath, which sounds like I am clearing my nose with a backward inhale (see p. 25). Horse friends approaching each other will make an array of sniffing, huffing, and nose-blowing noises before they even get close. By making these huffs and sounds, I tell Joe I am happy to see him, too.

Joe's high head, wide eyes, and sense of eagerness remind me of a child that is happy to see me. Joe nods his head, welcoming me into his Bubble of Personal Space. He is already reaching his nose out toward me as I walk to him. Everything in his body language is telling me he is eager to have me close to him. This is a far cry from our first meeting.

We sail through the Greeting Ritual, and he enjoys Copycat. He shows me he wants to follow me. On the third Knuckle Touch, he gently arches his neck and lips my hand and arm slowly and deliberately, saying, "I want to groom *you!*" I can't help but smile at his sweetness. I reach up toward his neck and withers and give him a satisfying scratch. I hold out my other hand in front of him and he lips it tenderly.

I get his halter and step inside his stall. I ask him to step back and yield his space to me. Then I ask him to step over toward the wall, using his Go Away Face Button and Shoulder Button with only my fingertips. He steps lightly and gently over and away from my space, holding his head straight and low. He is saying, "I am totally willing to do what you want. Do you see me holding *very, very* still? I am a good boy!" I parody his voice to his owners, who laugh and tell him out loud, "Yes, Joe, you are being a very good boy!" Joe seems to catch the humor, because he looks up at his owners and his ears flop to the side as he Yawns. He shares our good mood.

Now, I turn and press my back up against his shoulder and ask him to get into the Horse Hug position to put his halter on (see p. 99). Most people face a horse to halter him. There is nothing wrong with this, however, in Joe's case I know the halter causes him a bit of tension. I might instill a new attitude about the halter with this approach via the Horse Hug position. Joe drapes his head across my chest almost immediately and with only a tiny hesitation. I stroke his face, then let go, repeating the Hug a few more times: I stroke his neck and mane and up to his poll from where I am standing at his shoulder. Then I stroke his ears and between his ears to his forelock. I discovered that stroking all these areas from behind the head encourages the horse to let go of tension. Horses rub their necks on each other in these places, and foals do a great deal of neck rubbing in this way on their mothers.

Slipping the halter on causes Joe to tense up slightly, so I repeat the procedure a few more times until he is more relaxed. I explain to his owners that the more tension I can get out of him now, the more likely he will stay calm as we go for a walk. I use Inner Zero, breathing deeply and imagining a "smile inside." I want Joe to find his Zero, too.

Breakthroughs happen at Zero. Although there are many techniques, tips, and tricks that all have value, horses heal and move into a better place when they find their Inner Zero near us. Otherwise, training techniques just pull the rug over the dirt on the floor. The dirt is still there, you just can't see it. Leaving the issues inside a horse and layering training over the top of them does not work in the long run. You cannot force relaxation in the horse, but you can be at Zero within and hold that space so horses can join you there.

I want to present a calm inner state that encourages Joe to join me *before* we leave the safety of his stall. I take a few steps backward in an "O" Beckoning Posture. I welcome Joe into my space. As Joe follows me in a circle around his stall, I place my feet deliberately, anchoring each foot to the ground before I move the other foot. I'm asking him to Match Steps, which creates clarity about what my feet are doing (see p. 85). Horses follow the sounds of each other's feet. Horses that are emotionally unstable benefit from a peaceful herdmate that moves slowly and clearly.

As Joe follows me, he lowers his head as if to study each step I take. He mimics my stride, how high I lift my knee and the exact speed I move. Before I stop, I pivot, then clearly bring both feet side by side and Stomp to Stop (see p. 86). He brings his front feet directly up behind mine and halts. I have set up the maneuver so my back is now toward him with his shoulder close. I can invite his head over my own shoulder and move easily into a Horse Hug. I want Joe to have a completely different Conversation with me about walking together than any he has had before with a human.

We step out of his stall and into the aisle. Joe's attention shifts in anticipation of what is *out there.* I am not surprised, but I must interrupt this pattern immediately. I stop both my feet, and emphatically step into his space so that he must step backward. He is confused but matches my steps. As we begin to walk forward again, he pays closer attention to my feet and movement. To keep our time together light and interesting, I Pause mid-stride and suspend my foot in the air. He is not sure what to do, so after he places his front foot down, he lifts it up again and sort of swings it back and forth.

# IN HAND

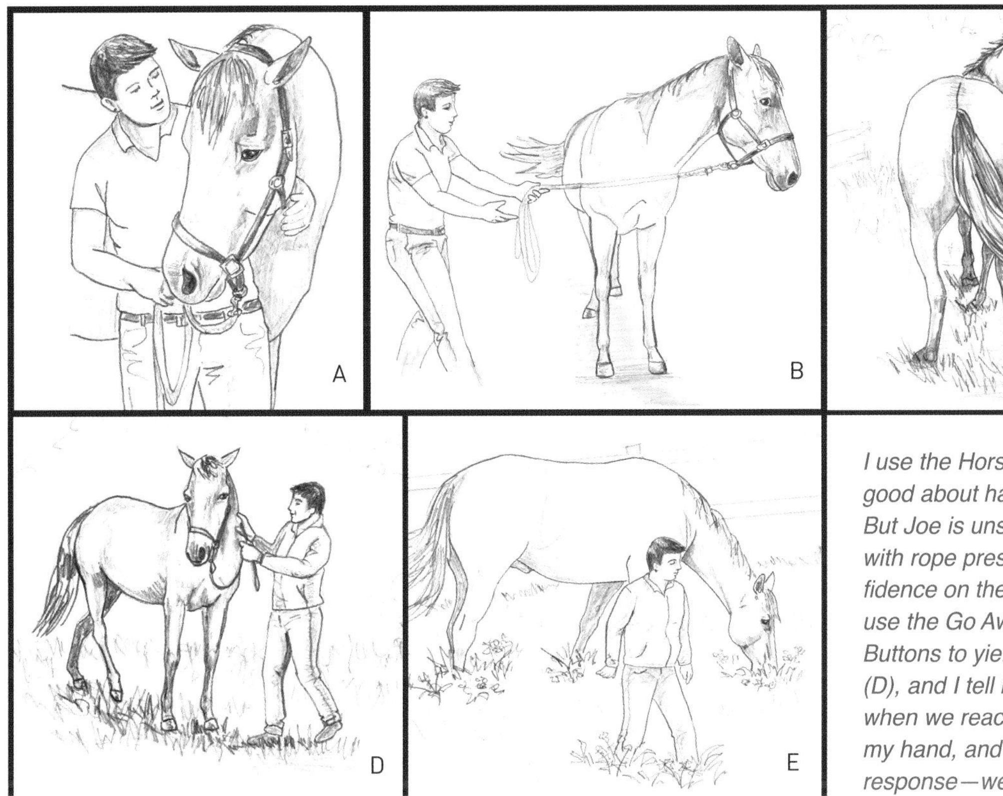

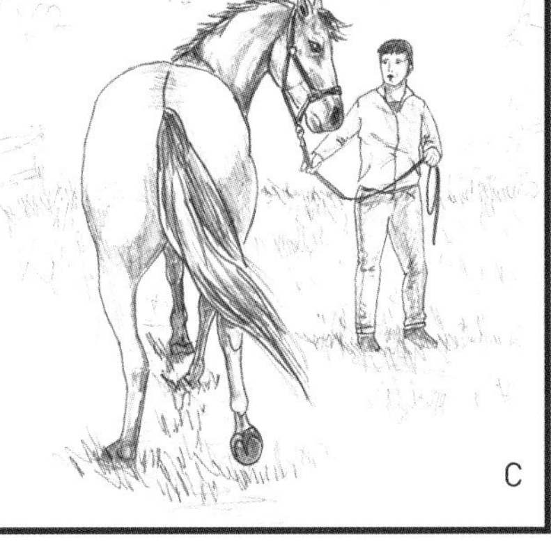

*I use the Horse Hug to help Joe feel good about having his halter on (A). But Joe is unsure about what to do with rope pressure and lacks confidence on the lead rope (B & C). I use the Go Away Face and Mid-Neck Buttons to yield Joe's front feet over (D), and I tell him "Good job, Joe!" when we reach his paddock. I swish my hand, and he swishes his tail in response—we are both done (E).*

Laughing, I place my foot down deliberately. Now Joe is realizing this could be a game. He likes games. I step backward, and so does he. I step forward three times then cross my feet sideways and so does he. I suspend my foot a few more times, and he tries to hold his up as well. His owners enjoy our antics. They have never seen this big horse so attentive to someone's feet, I call this Fun with Feet (see p. 88).

As we approach the barn door to outside, I feel Joe speed up, see his head go up, and notice his eyes begin to stare fixedly at what's beyond the door. I decide to have an Approach and Retreat Conversation about the barn door.

I make the interaction about Matching Steps, not about Going Somewhere in particular. This follows the protocol Joe's mother set for him. She would move off into the world only when she knew he would follow her feet. Young horses race around and check out all the surprising things in the world, then run back to their mothers. However, if Mom blows a Sentry Breath and raises her head, the games are off. Baby Joe would have stuck to her flank and kept his feet in exact time with his mother's when told to do so.

We begin our approach to the door again. I stop, lift my gaze up to the "bogeyman" in the far distance, and blow a Sentry Breath. This tells Joe that, for the moment, he should stick with me. Then I relax and pretend to chew gum. Now I can move forward through the door and outside with determination. I've let Joe know he is safer listening to me, and he is starting to believe me. He is a playful and thoughtful horse that worries about the big wide world. It may feel good to him to rely on me for his safety.

For now, Joe is following my feet and my ideas. Next, I introduce use of the lead rope. He still has a tendency to pull on it or resist it, which is his lifetime habit. I ask Joe for the Therapy Back-Up first (see p. 81). This puts him into a relaxed posture and makes him more open to the next step. I use the Go Away Face and Shoulder Buttons to ask him to step sideways. Although I have done it in his stall, this is the first time I've used it with him outside. Joe acts a bit "sticky" with the sideways request. He also puts his ears back and swishes his tail. A Tail Swish means, "No, I don't like this," or "Stop it!" (see p. 129).

I duly note his Tail Swish and his negative ears, but I also see that he makes an effort to comply. Joe is conflicted: he is enjoying the Conversation and feeling understood, and he is grateful that I am communicating with him. But...we have zoomed in on a problem area.

For whatever reason, Joe does not feel that he can completely let go and let me out-rank him. If he does, it means he must follow my lead no matter what. Essentially, he has allowed me to play the game of outranking

> **Friendships require the test of time and repetition to prove they are real.**

him, without fully committing himself to it.

I don't expect him to buy into my ideas right off the bat. Friendships require the test of time and repetition to prove they are real. All this means is that I have to repeat the sideways-yielding Conversation later. I coach his owners, suggesting they plan to have it with him dozens of times in predictable ways that are free of negative emotion. Like when coaxing a fearful child into the pool, we will all have to just stay open and positive until Joe sees, no matter what, we won't let him drown.

Since Joe liked Matching Steps and Fun with Feet, I use them for our walk together. I take a number of steps forward that are like steps in a marching band, then stop and change direction, Beckoning, then asking him to back up and yield his front feet side-ways. Every few minutes we Pause and rest. Each time we halt, I cup or scratch his neck. I acknowledge any bogeyman he sees and blow it away. I also pretend to chew gum.

Our Conversation sounds like this:

Me: "Joe, I want to walk three steps for-ward, then step to the side, and back up one step."

Joe: "Okay, but I want to run to my pas-ture so I can get to my next designated safe spot in this world."

Me: "I know, but you are doing very well. Let's stop and breathe and Share Space together out in the middle of the driveway. This spot can be a new safe spot."

Joe: "I am trying. This is hard, you know. There are bogeymen everywhere. I trust you, but I have never been able to tell anyone about how worried I get."

Me: "I know. As soon as you Match Steps with me without running away, pulling on the rope, or pulling back, we will go straight to your paddock."

Finally, we arrive at Joe's paddock, and as I open the gate, we have stayed in step with each other. I do not allow Joe to simply barge into the pen; instead I ask him to take the same tiny steps I make. Once we are in the paddock, I take a moment to Share Space with him before I turn him loose. I wait for Joe to lower his head and blow out his breath. We stand there in silence for a full moment of breathing before I reach over and slip off his halter.

Our Conversation has ended so I swish my "tail" at him by using my hand to swat once down by my leg. I turn slightly away from him and stomp one foot. Many of us have seen horses do this. In horse language I just said, "We are done Going Somewhere; our connection is over." It is important to remind Joe I'm still his leader. I govern how we close our activity together. Joe should not just run off and end our connection—I will do it. He swishes his tail as I leave to acknowledge this.

Joe moves off a little ways, then has a nice roll. His owners remark that usually when they turn him loose he runs off and "blows off steam." They always enjoy seeing his spunky energy. They ask why he is different this time. I explain that although his "spunk" seems like fun to watch, it is a sign of his insecurity. All horses will kick up their heels and play sometimes, but they should feel safe with us and should not begin the day running away.

Joe inspects his hay and water then comes to the fence where we are standing. He hangs his head and softly looks at us. His lips are quivering—a sign that he is feeling happy. Mike instinctively moves to him and rubs his face, reporting Joe never acts this way.

Mike and Liz are no longer labeling separate behaviors and feeling blinded to his genuine confusion about life. They can now see how seemingly random events are all connected and part of a bigger picture. They are also able to see why more or different training techniques have not helped with Joe. Most of all, they no longer feel like they are failing or are "bad" horse people, as though somehow Joe's problems were a reflection of their skills.

His owners express how awful it must have been for Joe to be so confused this whole time. Joe's issues were never tied to some elusive lack of training but were thoughts and feelings that were lost in translation to his humans. Animals have their own minds. They feel affection, pleasure, connection; they have wants and needs; they feel fear, pain, and protectiveness. We are all mammals and we have similar innate drives. Learning a horse's language puts us in touch with our own deeper sense of self. We live in such a speech-oriented society. Our own body language is "in there," and we loosely refer to it in certain gestures; however, I have found in my work with people and their horses that our body language has become sort of "vanilla." We have a generic association with a few gestures. Horses live in a world in which their thoughts and feelings can *only* be conveyed through their gestures, postures, and intensity levels. The same posture can mean two different things when one horse is breathing differently from the other. Observing the subtlety of their realm is quite a journey—one Joe's owners are just discovering.

Joe's personality indicates that he is immature. He wants constant reassurance about bogeymen and constant connection that is easy for him to read. Because of this, extreme footwork is going to be an essential part of helping him gain confidence. Horses that are nervous or seem ungrounded and flighty need to follow the feet of their leaders even more than other horses. Stoic horses may enjoy "marching" along and matching steps because this is very satisfying and important for the herd. But a stoic horse does not need Matching Steps in the way that Joe does.

When I halter Joe again in his paddock and walk back toward the gate, Joe is Matching Steps well. I hand him over to his owners and have them practice moving in time with him. It is important to move your feet in predictable and audible ways, but not to move too fast. Slow, loud steps are essential in helping a horse like Joe drown out all other distractions. His head gets lower as he begins to relax. (While it is true that we can teach horses to lower their heads, a lower head will naturally happen when the horse feels calmer and trusts the connection with the people around him.) Finally, Joe's owners "march" Joe back into his stall without too much fuss. He is making progress in letting go his life-long habit of being afraid.

I ask Mike and Liz if they have ever longed or round-penned Joe. They reply that he can be terrible on the longe, and they once took him to a round-pen clinic, but the next time they put him in their round pen at home, he tried to jump out. They often longe him before riding to "get his bucks out," even though he occasionally zooms around at top speed and is hard to stop. They are quick to add that when all is well, he can be a dream to ride. He is really sensitive and aware. He does not fight and loves to jump. If he spooks while under saddle, however, he has been known to take off bucking. Trail riding is difficult because he is so spooky; they try to take him out with other horses, which usually helps.

I am not going to train Joe to stop bucking or go on trail rides. Nor am I going to round pen or longe him in the ways most people might use these methods. My next step is indeed *in* the round pen, but I am not even going to be in the pen with him. I am going to turn him loose in there and talk to him about following my lead, and the lead of his people. Once Joe gets on board with the idea that humans can and will share the worries of the world with him and, in fact, help him to feel safe and relaxed, I believe Joe will do anything I ask.

I have Mike "march" Joe over to the round pen. Joe does fairly well; I can see the great effort he is putting into catching himself before he "freaks out," instead staying with his owner's steps. The wind picks up and a tin can falls and rolls near the back door of the barn, which is now behind Joe. He shoots right up in the air, but lands again, splay-legged and shivering as Mike stomps his feet and spontaneously blows a Sentry Breath toward where the noise came from. Joe immediately ducks his forehead into the area of Mike's chest and tries to take a big breath. Mike "chews gum" and strokes his big head tenderly. They just passed a test together, and both of them experience the dawning realization that things will be different now.

# Moving with Grace

## BECOMING MORE HORIZONTAL

We are vertical beings and horses are horizontal. In order to converse like a horse, we need to become longer on the horizontal plane like a horse. *Dancer's Arms* is a Conversation that provides my adaptation of how horses steer each other and the main skill you need in learning how to longe or work with your horse in the round pen.

■ One of your hands leads the horse. It is the target hand that replaces the head of another horse, walking in a parallel formation. You use both your left and right hand as the lead hand, depending which direction the head of your horse is facing.

■ Your other hand gives Sending cues at the Girth Button. Dancer's Arms clarifies how your Core Energy influences the horse. You cannot help but become aware of where your belly button is pointing when you hold your arms out from your shoulders. Your horse will completely understand how you are directing his body and face because you will begin to clearly show your cues to him with your body language.

## Keys to Horse Speak: Step 9

**Vertical vs. Horizontal (p. 141)**
**Dancer's Arms (p. 142)**
**Pivoting Energy (p. 144)**
**Lateral Movement (p. 146)**

I call this position Dancer's Arms to remind us to not be stiff and clamped in our body language. Instead we should be as soft and graceful as we can, like dancers performing onstage. The goal is to be light on our feet and ready to move.

Note: For some extremely sensitive horses, moving in Dancer's Arms may be too much pressure. Also, when first trying this, don't practice Dancer's Arms too close to a wall or a fence line. Some horses may feel squeezed or claustrophobic and try to scoot away from you because you are activating the Girth and Jump-Up Buttons. This Conversation is meant only to help you find out how your Core Energy opens or blocks the forward motion of the horse.

### *Conversation:* Dancer's Arms

**1** First, practice Dancer's Arms without a horse. Standing in a comfortable stance with your belly button facing straight out from your body, hold both arms relaxed but at shoulder height, palms down. Palms down softens your shoulders and sides of your body.

**2** Now, maintaining the same arm position, rotate your hips toward one hand so they are at a 45-degree angle to your arm. This position will allow you to walk forward, following one extended arm.

**3** Walk a few steps, then rotate your hips the other way and walk in the opposite direction. It feels like you are doing the tango!

**4** Try the movement with your palms up. Do you notice a change in your center of balance, if at all? This is a great opportunity to learn a little about yourself.

**5** Now add your horse to the dance. You're not trying to say too much in Dancer's Arms...yet. You are simply inviting the horse to move with you while you learn to manage your arms and Core Energy. It is best to begin to practice Dancer's Arms with your horse in a defined area, such as a wide, clear barn aisle or an arena, with two markers (plastic cones will do) set 10 to 12 feet apart. Your horse should be haltered and on a lead rope.

**6** Facing your horse on his right side, start with your right hand holding the halter lightly where the sidepiece connects with the noseband. Do not put your hand through the halter; just lightly hold it with your fingertips (figs. 9.1 A & B).

**7** Hold the gathered lead rope in your left hand toward the end of the rope. The left hand mirrors the pressure of another horse on the Girth Button (see p. 124 for more about this).

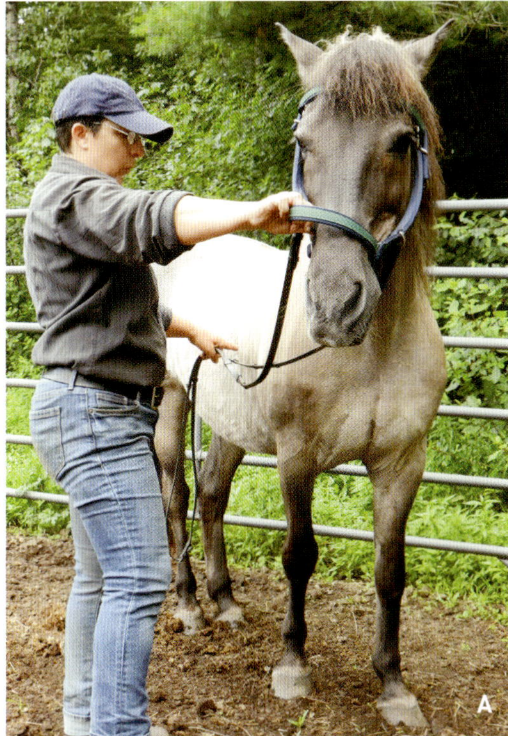

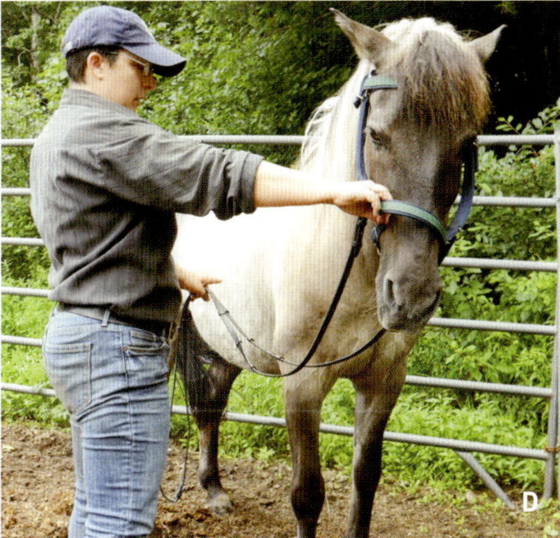

*9.1 A–D To start, get used to holding the halter in one hand, and pointing at the Girth Button with the other. I am not asking Rocky for forward movement yet—just getting used to the feel of Dancer's Arms (A & B). You should be able to motivate forward movement by pointing at the Girth Button (C & D).*

**8**  Use your left hand to cue your horse to go forward at the Girth Button, adding your finger, a swing of the lead rope, or the point of a crop if needed (figs. 9.1 C & D).

**9**  Imagine a flashlight beam coming out of your belly button—your Core Energy. Pivot your body as much as you can while looking at one of your markers (your destination). This will rotate your hips at least to a 45-degree angle toward the front

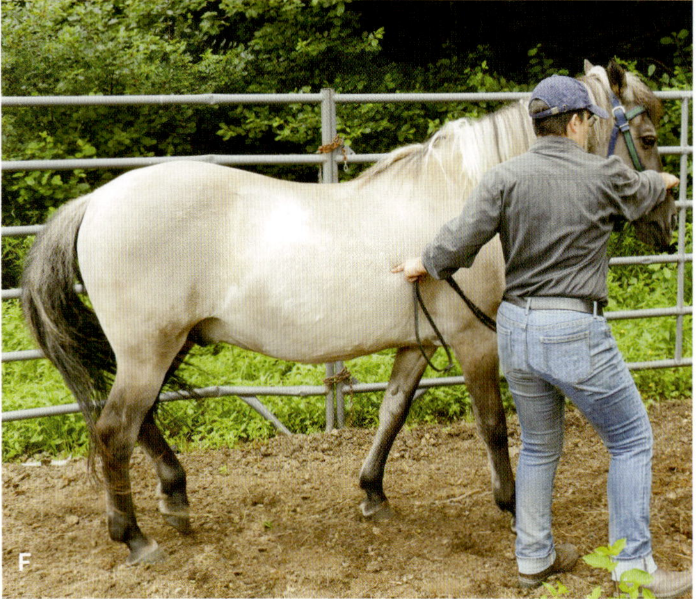

*9.1 E & F When I ask for forward steps, I use Matching Steps and aim my belly button forward (E). Now we are Matching Steps together (F).*

of the horse's chest. The angle of your Core Energy across and in front of the horse tells him to go forward, "But not too fast."

**10**  Try a few steps forward just to explore what happens with a "How Curious" attitude (figs. 9.1 E & F). You are saying to the horse, "Come with me, but keep your head straight. Do not go forward without me. Move forward but stay behind the invisible line my Core Energy is projecting across and in front of you, which will keep you collected in your body."

**11**  Practice Matching Steps and Fun with Feet, using Dancer's Arms.

**12**  Try this on each side of the horse and observe the difference. It might be more challenging for you to change sides than it is for him!

**13**  At any point where you or the horse is confused, Stomp to a Stop as in Matching Steps, but don't drop the rope. Blow out a breath and do Aw-Shucks.

**14**  Praise your horse for his try and giggle at your first attempt to "dance" with him—remember your sense of humor!

## Pivoting Energy

All horses can pivot each other using their head and hind-end gestures. Your arms and hands mimic these when turning a horse around with Dancer's Arms. As you shine your Core Energy in an arc in front of the horse, your arms follow. The horse's head will naturally come toward you in the pivot, but your target hand (the one on the halter at the beginning of the exercise) should prevent him from knocking into you. While you learn how to manage your Core Energy positions, it is fine to stop at any point in the turn by bending your knees and making an "O" Posture.

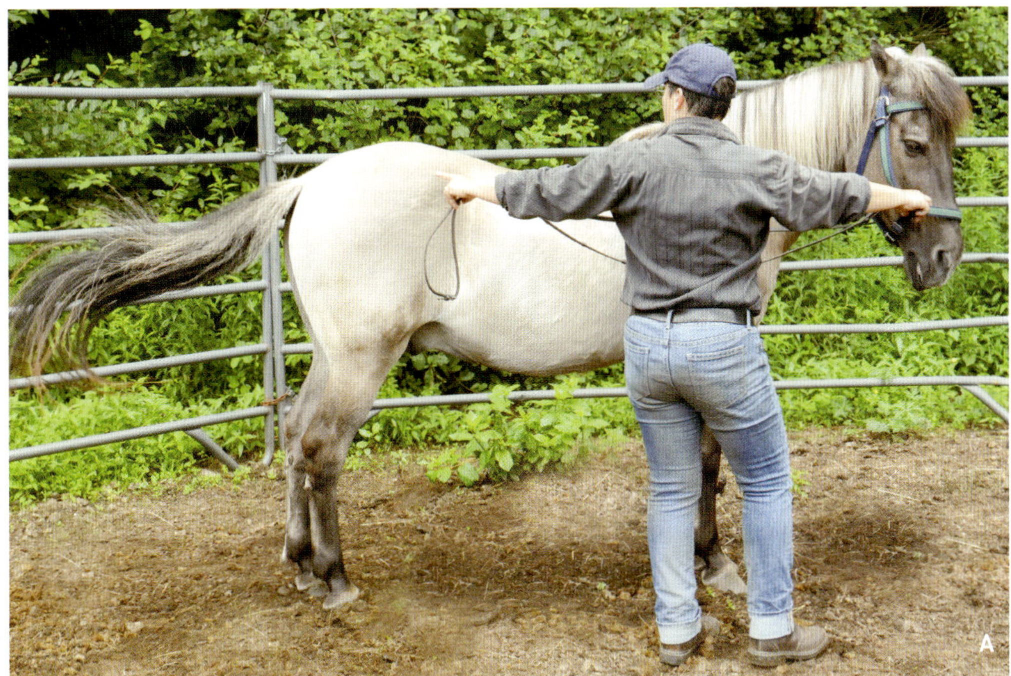

*9.2 A & B When my Dancer's Arms aim at the Hip-Drive Button as I arc my Core Energy in front of him, Rocky feels the message loud and clear, as his tail shows (A). We end with a nice Dancer's Arms pivot, changing direction (B).*

## *Conversation:* Pivoting with Dancer's Arms

**1** Place a cone or a marker on the ground to use as a target for your turn. Lead your horse in Dancer's Arms, but this time, prepare to shift your Core Energy as you approach the cone.

**2** Pivot your belly button "flashlight" in front of the horse, opening an arched path he can follow. Your Dancer's Arms may extend beyond the Girth Button to incorporate the Hip-Drive Button as you turn (fig. 9.2 A).

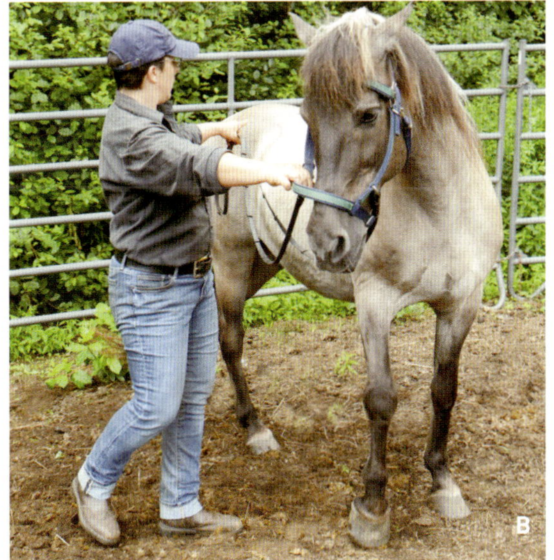

**3** The result is a change of direction around you and the cone (fig. 9.2 B).

**4** Try this on both sides.

## Lateral Movement

If you've had previous Conversations with your horse, he is now tuned in to your body language. So this time in the pivot, you can invite your horse to show you *lateral movement,* stepping sideways and forward at the same time. When you are proficient at forward Matching Steps and arcing turns with Dancer's Arms, *Lateral Yields* are easy. As we discussed on p. 124, horses yield lateral space to other horses from the Girth Button.

### *Conversation:* **Lateral Yield**

**❶**   Begin as you did for the Pivoting Conversation on p. 145. This time, just as you near the marker and begin to change direction on the arc, turn your belly button so your Core Energy shines right at his Girth Button.

**❷**   Step toward your horse and at the same moment, lower your center as if aiming your belly button *under* his girth line. You may also use your finger, crop, or rope to tickle this Button if needed. You may use one finger on the Go Away Face Button and the other on the Girth Button if the intensity level needs to increase.

**❸**   You are only asking that he *think* about yielding sideways as he moves forward on the turn. Praise him and yourself for whatever effort is made.

**❹**   Breathe out and bring both feet together, Stomp to a Stop.

Eventually, you may be able to ask for several sideways steps, or refine the exercise to begin shoulder-in and haunches-in in-hand.

# Who's Driving Anyway?

**W**atch one horse move another one off a pile of hay. Once the Sending horse has done this, he doesn't keep asking for that pile of hay. He has done what he needs to do. The mistake people make with horses is, we often keep pushing and repeating our cues—verbal and nonverbal—while the horse is moving.

We know horses Send each other away all the time; the horse that's been Sent away often hangs out on the outer edge of the Sending horse's Bubble. These Bubbles of Personal Space can be different, expanding and contracting depending on the situation and dynamic of the relationship.

- Here's how a horse stops Sending: He stops his language cues, turns his face or flank away, an Aw-Shucks, and perhaps some breath messages, or directs his Core Energy in front of the other horse and returns to Outer and Inner Zero.

- Here's what the Sent horse does: He moves in an arc around the Sending horse, and might even complete the circle, returning to his starting point. He often turns to face or look at the Sending horse to make sure the Conversation is over.

**Keys** to
Horse Speak: Step 10

**Clarity in Intensity Levels (p. 148)**

**Understanding Your Whip or Crop (p. 148)**

**Forward on the Longe (p. 152)**

**Trace My Bubble (p. 153)**

**Stop Pushing (p. 159)**

As you apply what you already know of Horse Speak to driving Conversations, the way both you and your horse think about leading, longeing, and round-penning will change forever. In these Conversations you will learn to pose questions such as, "How do you feel about my Sending you with your Hip-Drive Button, your Girth

Button or your Go Away Face Button?" You will ask your horse to Yield Space related to your Bubble in gradually increasing movements. You will assess how the horse answers each time, and therefore your response can be accurate before you ask for anything further.

In Horse Speak, we will pose these types of questions in three different settings. The reason is that our horses will have a different response to the pressure of our presence in each scenario.

- The first scenario is with the horse on the equivalent of a longe line, which we discuss in this chapter.

- Next, a similar Conversational template is applied with the horse at liberty inside a pen or paddock, while you are *outside* it, and the third setting is with both you and your horse at liberty together, *inside* the pen or paddock (we discuss these two scenarios in chapter 11—p. 160).

Horses view us as external members of their herd. This means Conversations about movement in Horse Speak will make sense to them. It's now time to learn to move your horse in more significant ways, the same way another horse would.

## CONVERSATION ON THE LONGE LINE

At this stage, longeing Conversations are all about finding clarity in your own intensity levels rather than "getting the horse to do something." It is your opportunity to learn to be clear in how you ask the horse to move as well as stop. Our usual confusing messages cause horses to resent activities like being longed—a useful tool that is too often misunderstood and misused. Fearful or resistant horses need this longeing Conversation to be slow and methodical to undo whatever negative experiences occurred in the past, while making your horse feel safe and encouraging him to become curious on the longe circle.

Before you begin your longeing Conversation, take all the time you need to make sure your horse is acquainted with the whip.

### Understanding Your Whip or Crop

The whip or crop you use for your first longe-line Conversations is basically an extension of your pointer finger. There are three positions by which to hold the whip:

*10.1 A–D In Walking Stick, the tip of the whip remains anchored to the ground so as to prevent unintentional cues and gestures that might confuse the horse (A). This is also the position used when showing Outer Zero (B). The Sword position is when you hold the whip straight out perpendicular to your body (C), and with Baton, your whip is parallel to the ground with your hand further down the handle or even in the middle so you can use either end to point at the horse, as necessary (D).*

- Position 1: Walking Stick—One end of the whip is anchored to the ground as it would be if you were using it like a walking stick (fig. 10.1 A). This prevents aimless wiggling of the whip, which only confuses the horse. This is also the position used when showing Outer Zero (fig. 10.1 B).

- Position 2: Sword—In the Sword position, hold the whip perpendicular to your body with straightened arms, aiming at a Button on the horse (fig. 10.1 C).

- Position 3: Baton—With the Baton, hold the whip down the handle and parallel to the ground (fig. 10.1 D). This position makes it possible to use either end to point at a Button at the front, middle, or back of the horse.

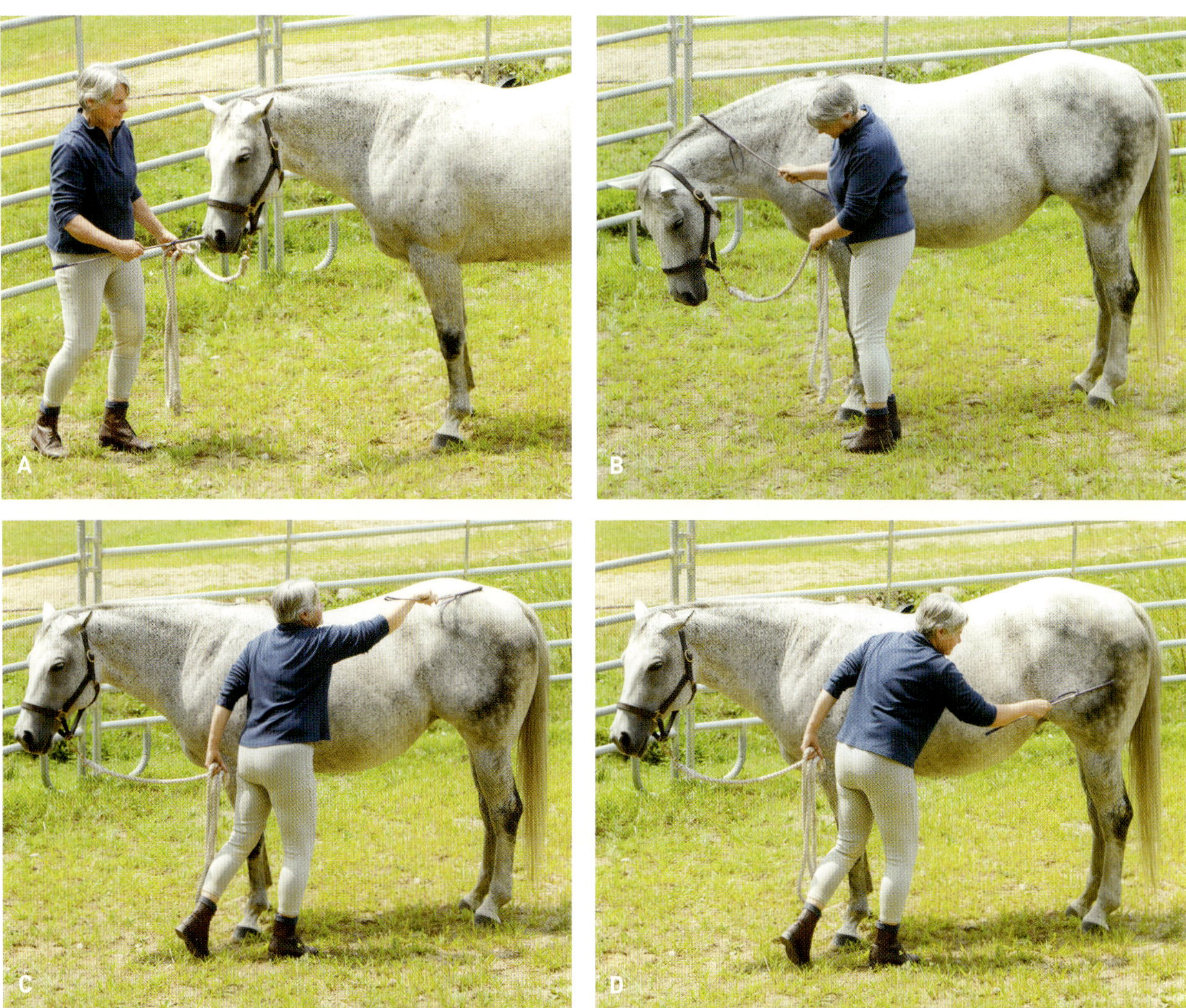

*10.2 A–D* *When your horse is nervous about the whip, start out using it in a Greeting Ritual (A). Most horses can learn to enjoy being stroked by the whip. Gretchen scratches Image's neck with the handle (B). Approach and Retreat all over the horse's body. Eventually, most horses learn to see the whip as a new Grooming tool (C & D).*

### *Conversation:* **Greeting Your Horse with the Whip**

**1** Keep what you learned during the Greeting Ritual in your mind as you first approach your horse with the whip (see p. 52). Let him sniff the handle first, as if it was your hand extended, knuckles up (fig. 10.2 A).

**2** When your horse is showing signs that he is not afraid of the whip, scratch him with the thick handle end, using an up-and-down motion on the neck near the mane where he might enjoy it (fig. 10.2 B).

**❸** Present the handle to his nostrils again, encouraging the horse to sniff it. You might even scratch around the edge of his nostrils with it or bend over and take sniffs of the handle yourself.

**❹** When your horse is still fearful of the whip, practice Approach and Retreat to help your horse begin to trust what can be a scary object (figs. 10.2 C & D). The ultimate goal is to be able to rub every inch of your horse with the whip and have him feel at ease.

### *Conversation:* **Practice with the Whip**

**❶** Hold the whip or crop in Walking Stick position, ready to be engaged at any time. When you are not using the whip, it should be in a low dragging position behind you. Avoid aimless wiggling, as the horse will assume you are saying something.

**❷** Try Level One intensity: Point the whip at a Button on the horse with intention—the equivalent of a small "X" (figs. 10.3 A & B).

**❸** Now take it up to Level Two intensity: Make a motion with the whip as an indicator or a tickle touch can be used here as well. Add vocal motivation, such as a kissing sound. Think: Medium "X."

**❹** At Level Three you are make a larger "X." Use the whip to "scoop up" energy from the ground *toward* a Button. Include a light touch or tap and vocal sounds as needed.

**❺** At Level Four, slap the ground with the whip, or stand in a jumping-jack "big 'X'" position, holding the whip as an extension of one arm.

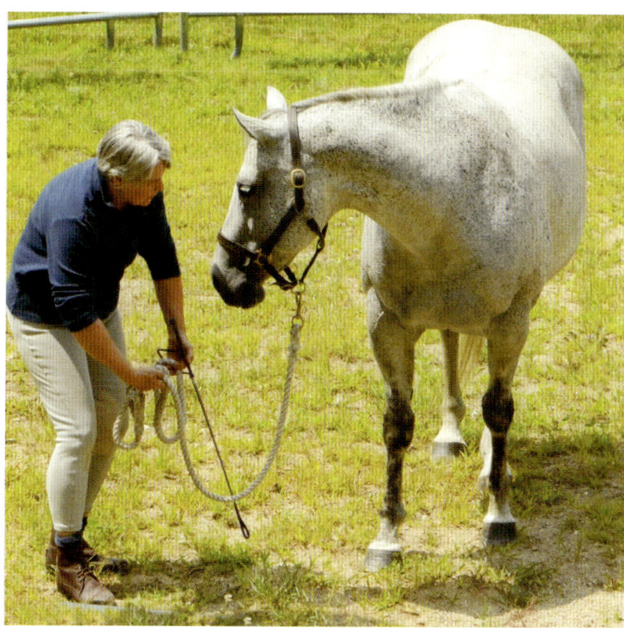

***10.3 A & B*** *Gretchen asks Image to move her face only, with a point of the whip and Level One intensity (A). She is welcomed back with the "O" Posture and the whip pointing down to the ground (B).*

Just like with the other Conversations and skills we've discussed, it is important to remain at Inner Zero at all times. When we need to raise our intensity, sometimes emotions or overzealousness will come up, and if you feel anxious or frustrated, for example, your horse will know right away.

## *Conversation:* **Forward on the Longe**

Longeing is actually asking a horse to move along the edge of your Bubble of Personal Space.

**1** Start with your horse only 3 to 5 feet away, knowing your arms and Core Energy will have the same Conversation as in Dancer's Arms (see p. 141). Angle your Core Energy toward the horse's Shoulder Button; this slight Sending pressure acts to keep the horse on the edge of your Bubble (fig. 10.4).

**2** Your leading or target hand is the one closest to the horse's head. It holds the 8- to 10-foot lead rope. I teach students to begin with a long lead rope rather than a longe line to reduce stress on both you and your horse. Handling a full longe line and longe whip is cumbersome; right now you need to be able to quickly change the Conversation if your horse is reactive.

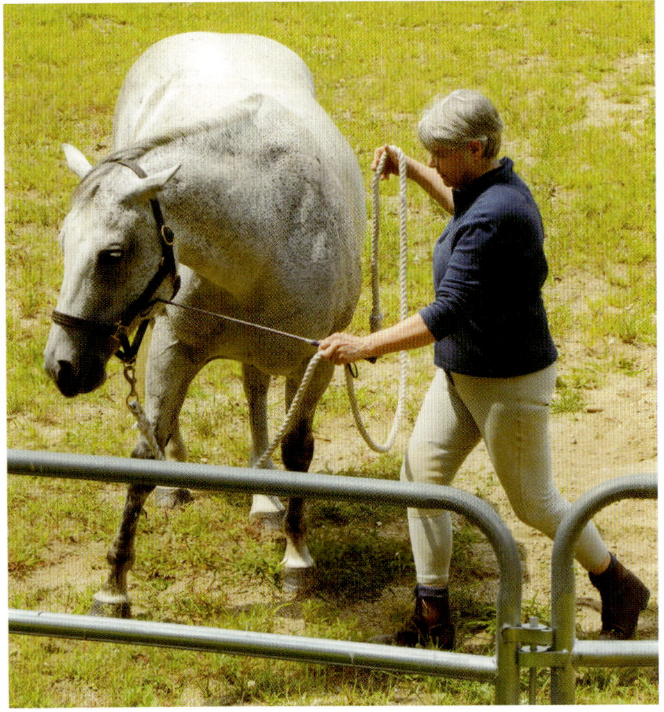

*10.4 Gretchen engages the Go Away Face and Shoulder Buttons to ask Image to step aside.*

**3** Your other hand holds a short whip or crop as an extension of your pointer finger. This hand also holds the extra length of lead rope. Holding Inner Zero and Dancer's Arms, point the crop at the Hip-Drive Button with a Level One or Two intensity. You've just asked, "May I move you forward using your Hip-Drive Button?"

When you begin, you are trying to build confidence in your horse. With this in mind, place yourself on the side that he finds most comfortable. Once he is thinking well, move to his other side. There is no mystery to creating acceptance and this whole Conversation allows you to practice talking to the Buttons, as well as using all the other tools of Horse Speak.

There can be no wrong answer from the horse. He might Turn the Canoe, facing you as his mother taught him to do. Or, he might not understand what you are asking. Some horses may not trust you to drive them or may be nervous about being driven forward. If you feel threatened, keep the horse's face away by directing the whip or crop at his Go Away Face Button. Be aware your horse might answer you: either by standing immobile with ears back, hard eyes, or tight lips, or by moving around you in a chaotic or frantic way.

No matter what he does, your reply should be the same: Drop all cues and take the pressure off with Aw-Shucks and breathing. This tells him you are saying, "Okay, no problem, thanks for answering me." As soon as his reply changes, reward him no matter what he does to get him to *think*. Reply back in a way you know he likes, whether it's breathing with him, scratching him with the whip handle, or doing Rock the Baby on his withers or halter. As we learn Horse Speak, we must always allow space for the learning curve for both ourselves and our horses. See if your horse can move one or two steps. Pause and go to Zero, and repeat three times in a row, rewarding each time you drop your cues and "Stop asking," no matter what his answer is.

Your goal? You are initiating a discussion, one that introduces the possibility that he will become open, interested, and curious about moving around the edge of your Bubble of Personal Space. In Horse Speak, when one horse asks another to move off, it is not uncommon for the driven horse to feel grumpy or defensive about it. You want to transform this Conversation from one where your horse feels scolded and sent away to one of playfulness and curiosity. When his curiosity is engaged, this Conversation will become full of possibilities for you both.

### *Conversation:* Trace My Bubble

Once your horse is willing to take just two steps while you are in Dancer's Arms, in longe position, you can expand your Bubble of Personal Space and lengthen the line.

**❶**  Allow more rope to flow out and indicate that your Bubble has expanded by pointing the whip at your horse's Girth Button, Shoulder Button, Mid-Neck Button, or Go Away Face Button, depending on what your horse responds to best (fig. 10.5 A).

**❷**  Let the line out in measures, but not so far that you and your horse can't "hear" each other anymore.

**❸** Place the whip against your belly button to make sure your Core Energy is aiming at the Girth Button so the circle you want your horse to travel is clear (fig. 10.5 B).

**❹** Flick the point of the whip between the Girth and Hip-Drive Buttons (fig. 10.5 C).

You want the horse's answer—whenever you ask—to be, "Sure, I'd love to move off upon your request." This will happen if you keep asking for forward and then stopping repeatedly, until he gets to a place where he realizes all you want is for him to walk and halt on the edge of your Bubble. Keep this from becoming drilling by frequently rewarding the horse. With repetition of ask for forward, stop, and reward, your horse will realize this is *all* he needs to do.

You may feel like an orchestra conductor, but as soon as your horse takes two steps, Drop It to Stop It, placing your whip in Walking Stick with the end dragging. Go up to him and reward both of you.

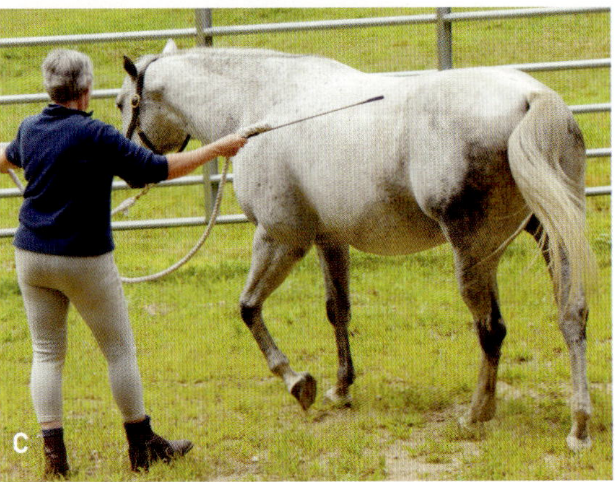

*10.5 A–C Gretchen moves Image forward on an arc using her target hand and the Girth Button (A). She places the whip on her belly button as Image arcs around her—this makes sure the shape of the half-circle is right (B). The pair reaches the goal of a relaxed half-circle and forward movement from the Hip-Drive Button (C).*

## Conversation: **Drop It to Stop It**

In most of the Conversations so far, you have been using Stomp to a Stop. Now you can graduate to a movement I call *Drop It to Stop It*. Before trying this on the longe circle, practice while reviewing a previous Conversation, such as Matching Steps.

❶ When you ask your horse to stop, Stomp to a Stop, but add a "sitting down" motion in your pelvis (fig. 10.6 A). By adding this body language, you are no longer talking to his forehand—you are now including the horse's pelvis. You are telling him to *stop through his whole body.* Note: Do not stop him until he has walked a half-circle around you. Most horses want to quit halfway around because when they are moving around another horse, it usually only takes an arc to readjust their positioning.

❷ At half the a circle, Drop It to Stop It, and make an "O" with your body, drawing the rope back toward you. Encourage the horse to, at the very least, look at you. You can also allow him to walk calmly back toward you if he shows interest.

❸ Go to him and Groom or breathe with him as a reward (fig. 10.6 B).

❹ Repeat the half-circle followed by reward and praise three times. Tell him how wonderful he is!

## Turn Previous Training into Thoughtful Conversation

By now your horse understands this is *not* his normal longeing routine. He is probably starting to "get" the concepts of moving forward and Drop It to Stop It. He appreciates having a chance to think and be rewarded with a scratch, perhaps Rock the Baby on the halter, and some praise. He is becoming curious about and invested in this Conversation.

*10.6 A & B Drop It to Stop It is an advanced use of the "O" Posture (A). Image and Gretchen have a lovely check-in as a reward after they stop (B). This is the first time they have tried using Horse Speak while doing groundwork.*

Horses will not make other horses do a complete circle around them, but as we touched on back on p. 147, often Sent horses will complete circles on their own. In addition, any horse trained to longe (and this is many a horse, since longeing is so often used in early training) knows you want him to do a whole circle around you. All you need to do now is turn previous training into thoughtful Conversation.

### *Conversation:* Completing the Circle

**1** From your Dancer's Arms position, ask your horse to move out on the lead rope.

**2** Stop using the whip the moment he moves—park it so it trails down behind you on the ground.

**3** To keep the momentum going without using the whip, march in place as if you were Matching Steps. Keep your "X" position and Dancer's Arms at the right level of intensity for your horse.

**4** Once one full circle is complete, stomp your feet, use your "O" Posture, and Drop It to Stop It.

**5** Praise your horse for a job well done.

**6** Repeat this Conversation three times to ensure your horse understands.

Remember, a handy way to make nice round circles is to place the handle of the whip against your belly button and aim the tip at the horse's Girth Button. This ensures you are not making an oval or aiming your Core Energy in front of the horse (which would tell the horse to stop and go at the same time). Being mindful of where your belly button is pointing will bring all your longeing Conversations to a whole new level.

### *Conversation:* Changing Direction

Once your horse walks calmly in full circles in one direction and you can halt him with an "O" Posture and Drop It to Stop It, take three steps backward and reverse his direction.

**1** Beckon him toward you with your "O" Posture.

**2** As he nears, switch your whip and rope to opposite hands and ask the opposite Go Away Face Button to turn and go in the other direction. Aim your belly button and Core Energy in front of your horse's chest to block his forward movement. This is how an Alpha horse can tell another horse to leave, stop, or change direction. Note: You may want to practice switching your rope and whip hands a few times *without* a horse, as it takes some coordination to master.

## Adding Obstacles

Horses enjoy longeing while weaving between cones, around barrels, and over ground poles at the walk. Incorporating obstacles helps you refine your "X" and "O" Postures, depending on the sensitivity of your horse. They also give you something more to talk about than just the longe line. You need to use the whip and the Buttons to ask his face to turn and look in the direction of specific objects to ask his front feet to move toward, over, and around. It's necessary to manage your Core Energy to ask him to step sideways, and your hand holding the lead rope must become a steady target for his eye to follow. Practice bringing your hand toward your belly button to ask his face to turn to you or lift your target hand higher to ask him to move toward an obstacle.

## Stepping Up the Pace

Horses run in a herd. Sometimes this is when there is a dangerous predator around or perhaps when they are being chased by bugs. Sometimes, they are simply playing up, kicking their heels, and running for the fun of it. For us to ask them to move faster, we must disassociate acceleration from why they might *run away* from something. You don't ever want the horse to feel chased by you. The first time you begin to ask for increased speed on the longe circle, he may believe you are telling him there is something to run away from, not: "Hey, let's move out a bit for fun."

### *Conversation:* **Trotting on the Longe Circle**

When you are ready to transition to the trot, repeat the same pattern you used at the walk (see p. 152), but now ask for more energy of movement.

**1** "Scoop up" energy from the ground with your whip acting as an extended pointer finger aimed at the Hip-Drive Button. This is your forward, Sending message.

**10.7 A & B**

*Gretchen jogs in place while aiming at the Hip-Drive Button, phasing up her intensity to invite Image to trot (A). Because trotting is so much faster, it is easy to get out of position. Place the handle of the whip on your belly button to check (B).*

**2** For initial clarity, you might choose to say, "Trot," out loud.

**3** Make sure you are Zero inside and use the minimum of intensity required. You may need to phase your intensity level up from Level One to Two to Three until your requests are clear. For example, an increased intensity might require jogging in place to encourage the horse to Match Steps (fig. 10.7 A).

**4** Remember, *any* answer gets a reward; don't have rigid expectations. This Conversation is meant for you to learn to calibrate your own Horse Speak appropriately. When your horse takes two trotting steps, Drop It to Stop It. Approach him and do whatever he likes the most, whether it is Grooming, breathing together, or Aw-Shucks and Scanning the Horizon. All you are trying to say is: "I just wanted to know if you could trot and stop. You can? Wonderful! Thank you!"

**5** Next, allow him to trot a half-circle. When he does, stop and reward with a scratch or breath communication.

**6** After three half-circles, ask him how he feels about trotting one complete circle. You have built a ladder of success in the previous Conversations starting with Dancer's Arms and ending with full trot circles on the longe. This is how you build up your horse's confidence. Allow him to talk about how lovely he looks when he trots! If you keep the fact that he might have something to say in mind, he will enjoy showing off instead of feeling merely obedient, coerced, and mechanical.

Remember, if at any time your horse becomes reactive, he has stopped thinking. The Conversation has ended. In his head, your horse is somewhere else other than the present. Perhaps he is reacting to a memory. Perhaps you inadvertently moved your whip in a way that confused him. Praise yourself for noticing the Conversation has just morphed into something else entirely, and if possible, go back to the first questions and answers you enjoyed at the walk *before* things unraveled. Alternatively, Greet your horse and just lead him while Matching Steps, which is the easiest version of Going Somewhere you can do together, or practice the Therapy Back-Up (fig. 10.8).

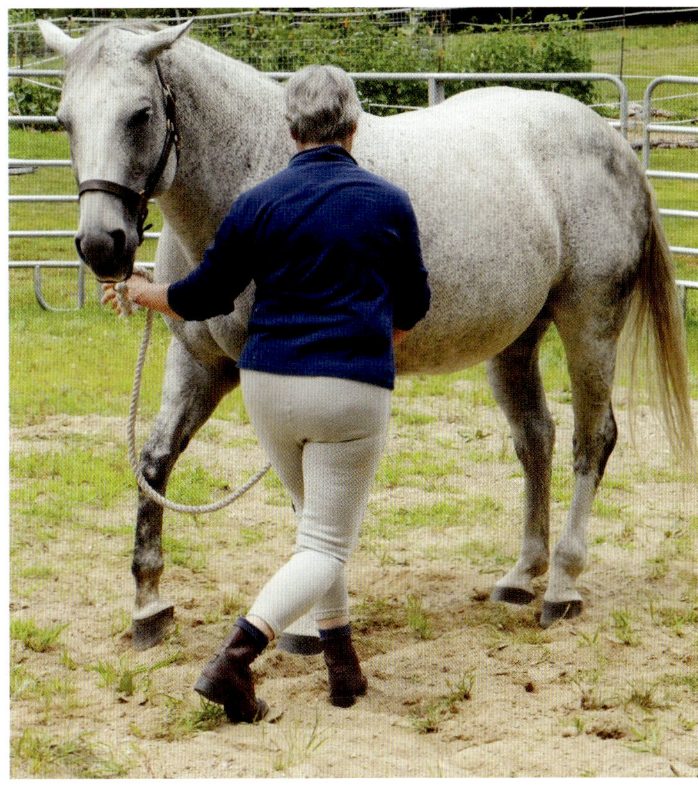

*10.8 Review other leading Conversations, like the Therapy Back-Up, if your horse gets reactive or confused, and to keep continuity.*

## THE NEW LONGEING YOU

If you, like most of us, were taught to push a horse around on a longe line, you will need to work at keeping your mind engaged in this new paradigm and *stop* pushing. Don't drive your horse just because you can. Chasing a horse around unfortunately taps into our predator instincts.

When you slowly build up to a point where your horse has a lovely balanced trot, let him keep it without continuing to cue him. Just allow him to move on his own. Hold your position, passively, in Dancer's Arms, watching for any language he shows you. If he goes from "listening" ears to flat-back ears, Drop It to Stop It, and rebuild trust with scratches, breath, or Rock the Baby. Praise your horse for being beautiful and for trying to accept you. Practice being a coach who believes his team can win, rather than a dictator demanding perfection.

Your intention is to create a soft, curious, and interesting Conversation about moving around your Bubble of Personal Space. A confident horse may circle you on the longe within minutes, but I have met and restarted a hundred horses who really resented this request at first. Using the Horse Speak skills you have learned up to now should set you both up for success.

## HORSE SPEAK AT LIBERTY

Our second and third Conversational templates are at liberty. Horse Speak in these scenarios furthers your ability to be assertive *without* becoming aggressive. The power of liberty work is that you are enhancing proprioceptive awareness to the tiny shifts in balance, tension, and pressure in your body. Your body moves and shifts ever so slightly, which we may not notice, but to a horse, this language is loud and clear. We gain much from a heightened sensitivity to our sense of our body's position, movement, and speed.

It should be noted that some horses are so reactive that Conversation is not possible when you are in the same enclosure with them. If this is the case with your horse, there is still a safe and totally effective way to have a liberty Conversation with him. I often have my students begin speaking to their horses at liberty from *outside* the fence, be it a turnout paddock, riding ring, or round pen. The horse—on the inside—is loose so he can think, relax, and be authentic. You can move along the edge of the fence line and observe how your horse looks, acts, and feels in response to your presence while he is totally free.

Learning the Conversation from the outside first ensures that, should the horse become excited for any reason, both of you are safe. You should not practice liberty inside an enclosure until you have calm and complete mutual *acceptance*, fostered through practice of Horse

## Keys to Horse Speak: Step 11

**Observe and Mirror (p. 161)**
**Over the Fence (p. 161)**
**Vocal Direction (p. 163)**
**Define Level Four Intensity (p. 165)**
**Parallel Tracks (p. 171)**
**Target Practice (p. 171)**
**Transitions (p. 174)**

Speak in the basic Conversations we've covered in the stall, in-hand, and on the longe. Even if you have progressed through this book from the beginning and have practiced Horse Speak in its various parts in different settings, both with a halter and lead and with your horse loose, there is still something to be learned from linking together the elements from the other side of the fence first.

## Observe and Mirror

Observe your horse once he is loose in the enclosure. Is he Scanning the Horizon, blowing Sentry Breath? Is he sniffing the ground in Aw-Shucks? Does he roll? Practice being mindfully present in order to observe him and what he is offering. Think to yourself, or say out loud, "You are loose and so am I. I am not attached to you with a rope. You are not expected to do anything because you are on the other side of this fence. I just want to know how you think and feel about us being together like this."

If you feel unsure of what to do, just start by being your horse's student. Follow all his movements. Mirror everything he does: Look where he looks, lift and drop your head as he does. If he puts his nose to the ground, do Aw-Shucks. Once you are comfortable in this dynamic, explore the following Conversation.

### *Conversation:* **Over the Fence**

**❶** Begin by calmly approaching the round pen or paddock in Matching Steps and releasing the horse within (fig. 11.1 A).

**❷** Stand outside the fence, but near the gate. This gives your horse incentive to come to you. Combine an "O" Posture Beckoning gesture with an Aw-Shucks scuffing of the ground. Most horses find this irresistible (fig. 11.1 B).

*11.1 A & B Entering the round pen should not be stressful. Use Conversations you've learned to make it calm (A). Beginning liberty work with you outside the fence gives both of you space to feel safe (B).*

**❸** Extend one hand through the rails and complete all three Knuckle Touches of the Greeting Ritual if you can. Reach in and Groom your horse with your fingers or the handle of the whip if you have made it a friendly tool already (fig. 11.1 C). If your horse walks away, swishes his tail, and avoids you, hang in there. He is *actively ignoring* you, which means you are very much on his radar. In this situation, simply mosey along the outside of the fence until you get in front of his face, reach through, and offer a Greeting again.

**11.1 C** *Image and Gretchen Share Space and enjoy the Grooming Ritual (C).*

When you get to the third Knuckle Touch, pay attention to what your horse wants in this new and different situation. Does he want Grooming or is he Gone again, turning to leave? Does he want to Share Space with you? If he is putting his nose on the ground—an Aw-Shucks—you need to answer with your own scuffling Aw-Shucks. Take big breaths, Scan the Horizon, maybe even blow a Sentry Breath at the far end of the pen. Turn your back to him for a minute or so in a Pause so he can think.



**❹** Once you have had a satisfying Greeting and have Shared Space or done some Grooming, you should have his interest and can move on to Going Somewhere. For this you may need a small crop or dressage whip to act as a finger extension, because you are on the outside of the pen. Keep the whip in the Walking Stick position at all times unless you are using it to Groom or point at a Button. Unconsciously wiggling the whip around will ruin the Conversation because the horse will be distracted by your confusing signals.

**❺** The initial Going Somewhere Conversation can be Fun with Feet (see p. 88). Use an "O" to draw your horse's body to you again. If he does not want to come, combine "O" with scuffing the ground in Aw-Shucks at the same time. Smile

## Connection
### IN PRACTICE

Once, I spent an entire afternoon sliding back and forth along the outside of a round pen looking for connection with a horse that was actively avoiding me. My persistence and presence paid off. Eventually, my mirroring actions made him so curious that he came over and did a perfunctory Greeting with me. When he lingered in my space after the third Knuckle Touch, we were on our way.

*11.1 D–F Gretchen moves down the fence and Beckons to Image (D), who enjoys coming back to the "O" Posture (E). Gretchen then asks Image to Go Somewhere with her using the Hip-Drive Button (F).*

and Beckon him with a lowered head and offer your knuckles (figs. 11.1 D & E).

**6** Once he is paying attention to you, stand a little way back from the fence and pick which front hoof you will move. Lean in, looking at it, point the whip, and deliberately take one step toward your horse with your foot. This game should already be familiar to you both. You may have to hold your foot up longer and step forward with exaggeration. The moment you see him thinking about moving his own foot, quit what you are doing, praise him, and

## Vocal Direction
### IN PRACTICE

In liberty Conversation, you can say what you are asking your horse to do out loud. While of course your horse may not understand the actual words, saying, "Give me your foot," causes your body to shift in micro-movements, which the horse *will* pick up on. As mentioned, liberty work enhances proprioception—our awareness of tiny shifts in our bodies' balance, tension, and pressure. When you speak out loud, your body will move ever so slightly, and your horse will "hear" your body movement. I have often helped horses and humans through misunderstandings by having students state out loud exactly what their thoughts and feelings are.

try again, at least three times. Many horses think long and hard. Eventually, he will offer to move his foot more quickly.

**7** Now, relax the whip into Walking Stick position and take a big step back with the same foot. Reinforce the attempt to Draw his foot toward you by making a big "O" shape with your body. With the hand on the same side of your body, create a pulling motion from his foot as though you had an invisible string on it. Once he brings his foot forward, and you have praised or rewarded him, step forward again, into his Bubble, asking him to step backward. Try the other hoof and your other foot.

**8** Once you have had Fun with Feet, move up to the horse's chest and ask him to back up a little by tickling the whip at the Shoulder Button.

**9** If he does not yield, position yourself at a 45-degree angle to his face and explore his Go Away Face Button. Wriggle your fingers in the direction of this Button. He should be familiar with this space-yielding request by now. Point at the Mid-Neck or Shoulder Buttons with the end of your whip, if need be. Now you've asked for space that includes him moving his feet.

**10** As soon as he yields at all, Beckon him back with an "O," take a deep breath or do Aw-Shucks, and praise him lavishly. You are thanking him for talking to you when he doesn't have to. Considering you are on the outside of the fence, all this may seem awkward. Although there is no wrong move in a liberty Conversation, consider your intensity levels. When he gives you space, adjust your levels and Beckon him back with a lower intensity.

**11** When your horse does not move off more than a few feet and you can see his Hip-Drive Button, he may be inviting you to drive him forward. I often find that thoughtful horses that are interested in conversing *do* actually offer the Hip-Drive Button on their own. You will know it is an "offer" because his body language will look like he is waiting for *you* to act. If your horse does not present you his Hip-Drive Button, you may have to mosey around the fence line of the pen until you are within sight of it. Again, you focus on this Button with the lowest possible level of intensity, and when you see a flinch or response, stop all cues immediately.

**12** This is where awareness of your "X" and "O" is essential, because here it has the most influence. Start with intention and stand up straight. Point the whip in a Sending message at the Hip-Drive Button. You will be making an incremental "X"—make sure you start with the smallest "X" possible and look for any reaction at all from the horse (fig. 11.1 F).

This driving Conversation, like others we've discussed, is important to your relationship with your horse. A very astute, confident, sensitive horse will be watching and will move off at the slightest intention toward his Hip-Drive Button. His reaction to your posture is the best indicator of how high on the intensity meter you need to go. Keep Zero intensity inside and stay open to what your horse says next.

**13** After you point at the Hip-Drive Button with the level of "X" Sending message necessary to cause the horse to move off, immediately return to an "O" Posture and welcome him to Come Back. Repeat the Go Away and Come Back three times, even if you have to change where you are located outside the pen. You can add a pile of hay near the fence so there is something to ask him to move away from

## High Intensity
### IN PRACTICE

Many horses have come to assume that *anything* we do is at Level Four intensity and react accordingly. You need to know where you are on *your own scale*. Once you do, you can make sure the horse is not just reacting to some old memory but actively "listening" to your request.

If you have a reactive horse, sometimes it is helpful to demonstrate what your Level Four looks like to him—that is, without frightening him and making matters worse. I recommend turning away from the fence and your horse and making a *huge* movement in the opposite direction. Pick an object to "get intense" with, such as a bucket or wheelbarrow. You will accomplish two things: First, the horse will see you scaring away an invisible threat. He may run or spook, but since your back is toward him, he will watch you to see what you do next because your energy

is not aimed at him directly. He will not take your Level Four intensity personally. Second, you show him how big and assertive you can be. As we know, horses respect assertiveness rather than aggressiveness.

In a herd, members must test their leaders sometimes to ensure that the Alphas are feeling well and up to the challenge of leading and protecting. A momentary demonstration of Level Four intensity is okay, as long as you stay at Inner Zero and as long as you are not feeling anger or frustration. The moment your insides start to churn with frustration, fear, or mental stories that have nothing to do with the present reality, is the moment you have switched from a healthy, assertive energy to an unhealthy, aggressive one. Your horse may test you for this possibility. Our incongruence in communicating can wreak havoc and confusion in our relationships with our horses. What horses feel on the inside is what they show on the outside, period. We have to make sure our insides match our outsides, too.

and come back to, adding a motivating value to the Conversation, if you like (see sidebar—p. 170).

## A Delicate Balance

Conversation with the Hip-Drive Button is finding the delicate balance between being assertive enough to cause your horse to move off, yet inwardly peaceful enough to make him want to come back. You've practiced this many times, but at liberty you get very honest answers from the horse. I do not recommend moving up to asking for trot or moving inside the pen until this simple exploration is well established.

When you have a horse that is clear, interested, focused, and eager to keep talking to you, and you feel that asking for a trot would be an interesting topic to discuss, you may proceed to the next step.

### *Conversation:* Liberty Trotwork

**❶** First, you must get good at turning your horse's head away from the outside of the pen, yielding the space to you so he is on an arc, facing toward the *inside* of the pen. If he is right up against the fence with you on the outside, ask his front feet to move off by aiming your Core Energy at the Shoulder Button, or wriggle your fingers at the Girth Button so he has clear direction to *move*. Ask him to *keep* moving by "scooping up" energy toward the Hip-Drive Button. Apply Sending messages to the Go Away Face Button, Mid-Neck Button, Girth Button, and Hip-Drive Button, in that order. Together, these express *one complete Sending message*. You are claiming the space on the side of the horse, as well as the space behind him.

**❷** Use Dancer's Arms like you did back on p. 157 to "scoop up" energy toward his Girth Button. What happens next depends on the horse. Some will keep moving at a fast walk, some will jump up into a frantic gallop, some will shut down and simply stare at you...and some will actually trot. No matter what he does, be consistent, remain at Inner Zero, and Pause often. When he trots, do a low jog-in-place motion with your own feet where you are standing, but facing the direction he is moving toward. Match Steps in line with his shoulder no matter how far away. Do not face your Core Energy at his Hip-Drive Button while you jog in step with him. This may be too much pressure and he may kick out in defense.

**❸** Ask again with the same sequence of Buttons, and when you get any response at all, stop asking and say thank you: Go to an "O" Posture and ask him to Come Back. Do a great deal of deep breathing and Aw-Shucks to tell him in his language that you are pleased he heard what you said and tried to reply. Extend your arm through the fence for a Knuckle Touch, repeating three times.

## Releasing Old News

Sometimes, after the second or third trot sequence, a horse might need to stop and "go inward." He may roll, lie down and sleep, or stand very still and blink his eyes. He may even perform the deepest level of releasing old news, which is to Yawn so hugely that his eyes roll backward. If you see that he needs a "time out" to process what you've done so far, then either sit down and go to Zero, or go away and come back later.

When you come back, your horse may look totally different: His eyes may seem

### "Calm Down, You Are Safe" IN PRACTICE

Remember, when Sending using the Hip-Drive Button, you tell the horse you will watch his back (literally). You must keep this in mind so you understand his various reactions. You are saying you believe you are responsible enough to protect him as you drive him forward from behind. He may say, "Yeah, right," or "No one else has, why would you?" Or, "You may say that now, but in a pinch you will make me do something I don't trust."

Especially in the case of a horse that does not trust people, asking him to trot might be the last straw, and he will implode, and then explode. If he *does* explode, do all the "Calm down, you are safe" Horse Speak sequences until you get his focus back on you and he has reentered the Conversation. Make an "O," do Aw-Shucks, Blow a Sentry Breath, Scan the Horizon, maybe even turn your back and/or sit down to really give him space to Pause and think. Try saying things out loud, like, "I know you think I am picking on you, but I really am not. I want to see your beautiful trot because you look so lovely doing it. I want you to feel happy about showing it to me. I want to help you let go of all this confusion, so I am going to sit here and stay calm so I can prove it to you."

Keep yourself at Zero and once he is calm, start over, Beckoning with "O" Posture so you can do the Greeting Ritual through the fence. Watch him closely, reading his face and tail as he sorts through his own inner story about what he thinks is happening. Repeat the Conversations you've had with success—Grooming, Sharing Space, Fun with Feet—until you arrive again at the Hip-Drive Button and trot Conversation.

brighter and softer, his mouth may be more droopy, and his shoulders may seem lower than before, giving him the appearance of somehow being a shorter horse. He may want to pick up exactly where you left off and whip through a very fast Greeting before promptly offering you the Hip-Drive Button.

This time, when you ask for trot, he may actually just trot along nicely inside the pen until you stop asking, go to "O" Posture, and invite him back. If he stops immediately, turns to face you, and strolls right over, tell him out loud how you feel. You both will probably feel very good indeed. I recommend having this Conversation to a peaceful conclusion three or more times before you even consider moving inside the pen with him. When you develop clarity of gesture and mastery of your phases of intensity, your horse will stay present and thoughtful when he is driven.

## CONVERSATIONS ON THE INSIDE

Being on the inside of the pen or paddock with your horse can be the most conflicting, confusing, and counterintuitive aspect of talking to horses. Driving a horse forward is very tricky to do without accidently triggering our own predatory desire to chase things or conversely feeling overwhelmed and even intimidated by being in a small enclosure with a loose horse. The horse can be a little or a lot more defensive in this situation because of the overlap of personal space, and we, too, may be defensive without the obvious safety net of a fence to stay behind or a halter and lead to depend on as so many of us are accustomed.

We've talked about Bubbles of Personal Space and the world of circles and arcs that horses move on and think about all the time. Both are hugely in play when working with your horse at liberty. As you will see, the horse can shift away or toward you, but it will still be in a predictable arc. Paying attention to where your belly button aims your Core Energy is more important than ever, now.

Having said all this, Conversations at liberty have a wonderful quality because the horse is truly free to express himself. Your previous Conversations in hand, on the longe, and from outside the pen have resulted in mutual *trust, respect, unity,* and *acceptance*. You both are more aware of how your "X" Sending messages and your "O" Beckoning messages feel and what each other's levels of intensities look like. You've both learned to stay at Zero on the inside and have volume control when it comes to intensity of movement. Every aspect of Horse Speak comes into play when you enter the round pen or other safe, enclosed space.

As you enter the pen, observe your horse for signs of stress. You now know what his tension looks like, so watch his face carefully and his tail, too, as well as how he is holding his body. I find that *many* horses have had negative experiences in the round pen by the time I am hired to work with them. A horse will tell you if this has been the case by his body language. If he rushes the gate or pushes into your space, he is feeling overwhelmed about something. Keep yourself safe. Go all the way back to leading, Matching Steps, and Longeing to build up to liberty practice in this space that causes him anxiety. I believe that horses remember experiences in "maps"—they do not generalize experiences until the same thing happens on at least three separate occasions. In the wild, if a herd gets attacked by a lion at one waterhole, the herd will avoid *that* waterhole, but *not all* waterholes. If a horse has a single bad experience in one round pen, he may want to avoid *that* pen, but be fine in another. But, if he had a hard time in a round pen every day for a month, he may associate all round pens as bad. When the horse becomes too tense when you enter the pen, drop back to the in-hand Conversations (see p. 78) and progress up to the ones done from the outside of the fence (see p. 160). This will reassure him that you are still just talking to each other. Your goal is complete calm when you are together inside *any* enclosure with him.

### *Conversation:* Beckoning and Sending from the Inside

**❶** Without holding any whip or rope, enter the pen and either Beckon your horse to you with an "O" or walk up to him and offer a formal Greeting to remind him you are there for a Conversation (fig. 11.2 A). If he wants to avoid your knuckles, Greet him with some deep breaths, then reach to softly touch his nose with your knuckles at least twice (fig. 11.2 B). This tells him that you are stepping in as his mother right

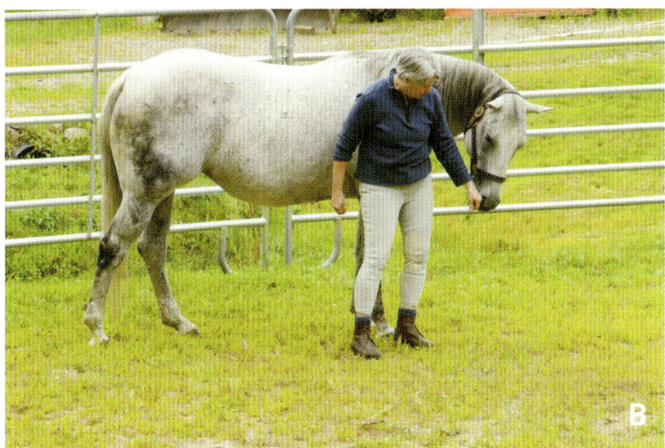

*11.2 A–C Since everything went well with Gretchen on the outside of the pen, she and Image will try a liberty Conversation on the inside. Gretchen enters and Beckons with "O" (A), then does her Greeting at a 45-degree angle to keep the pressure off (B). Standing to the side in "O," she Beckons Image to turn to her (C).*

now and she wants to talk. Your horse should relax because you are being a little assertive, but staying within the protocol. If you get three Knuckle Touches, scratch his withers, and be mindful to keep observing his gestures, as well as breathing. When you have an overeager horse, use the Go Away Face Button to define your own space. By now, you should know his favorite "calm-down activity," so offer that (fig. 11.2 C).

**2** Review the in-hand Conversations, now at liberty and if he's willing, linked together into a sequence. Start with Approach and Retreat. Step toward your horse and watch for a lift of his head, a blink, or any flinch. Stop and Pause, breathe, and do Aw-Shucks, which is your retreat. If he has moved off to a different part of the pen, it will be easier to see when you've reached the edge of his Bubble of Personal Space. Watch ears, eyes, and tail. Reward him every time he answers you.

**3** Now add the Come Back and Go Away Conversation. Imagine the direction your horse would move on an arc from the position of his head. Place yourself in front of him, on his arc, a similar distance from where you determine the edge of his Bubble to be. Make an exaggerated "O" along with an Aw-Shucks, and even call his name softly. If he even so much as leans toward you, praise his try.

**4** Now use a shooing motion, wriggling your fingers toward his Hip-Drive Button, Go Away Face Button, or Shoulder Button with just enough pressure to get some reaction. Again, if you get a lean or a flinch, he has heard you.

The horse doesn't literally have to come to you or move away for this Conversation to be a success. Repeat Approach and Retreat, as well as the Come Back and Go Away three times, depending on which one you both have more confidence in, and see how or if your horse answers differently each time.

## Predictable Directions

The following Conversations with you inside the pen will further inform you about your horse's arcs, curves, and circles. You will begin to see a predictable direction for his movements.

## Props
### IN PRACTICE

When a horse is nervous or stoic when I enter the round pen or paddock, I might put three small piles of hay on the ground and then move the horse with the smallest of Sending messages from one pile to another. I follow the Sending message with a step back off the hay and "O" Posture to Beckon him back to the pile I Sent him from. I may also offer to hand feed him a few pieces to calm him—horses do this with each other.

You can also begin Conversations with the Hip-Drive Button using hay as a prop, slowly moving the horse from one pile to the next, as long as he remains calm. Only attempt this if you have had success with "hay games," as I like to call them, from outside of the pen first, to be on the safe side.

*Conversation:* **Parallel Tracks**

**1** Make sure there is space on the opposite side of your horse and imagine two arcs, parallel, like the two rails on a railroad track: one that he is on and one just to the other side of him. Set your intention and ask your horse to yield just his head at first by indicating the Go Away Face Button.

**2** Next ask for one step with the front feet off his arc to the one parallel, using the Hip-Drive Button. The message is for him to move his head and take his feet with him. When he leans or steps at all, reassure him you *do not* want him to leave by immediately doing Drop It to Stop It.

**3** Now, without Beckoning him back to you, ask him to step over with his hind end to the parallel track. Use Dancer's Arms to support his positioning. Wriggle your fingers or touch lightly in the hollow of the rump just above the stifle at the Yield-Over Button (fig. 11.3). This Button is below the Hip-Drive Button, as we learned on p. 134. You might need to press your fingers into this spot until he moves one step over. He may Turn the Canoe and face you, or he might think you want him to move off and so walk away. As with everything else: You asked for something, you got something, so quit asking, and say thank you. You can iron out the details later. Your goal, however, is to place him on the parallel arc.

These dances of personal space and arcs and circles may seem confusing at first, but once you feel it for the first time, you'll be able to ask your horse to move to the parallel track with more clarity. Remember to have a playful "How curious" attitude and not be attached to the outcome.

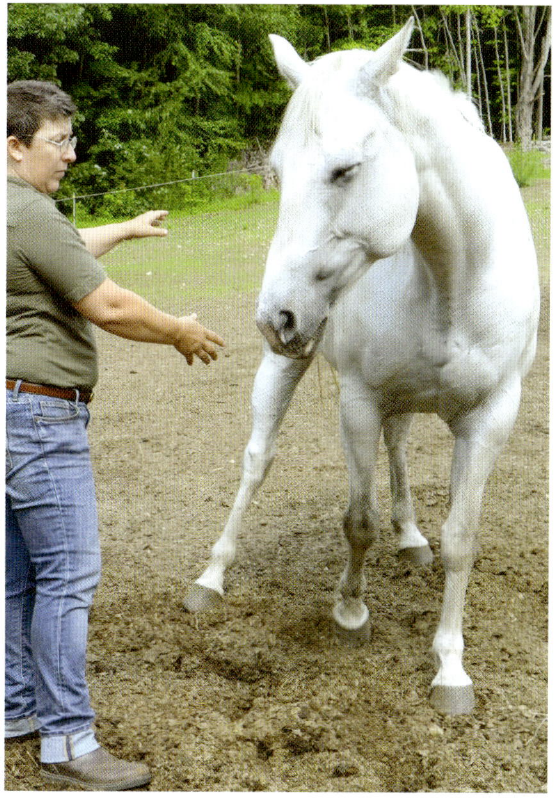

*11.3 Practice yielding the hind end over to the parallel arc or "track."*

## Target Practice

You should be able to use Dancer's Arms, your target hand, Matching Steps, Fun with Feet, and the Follow Me Button to initiate any number of games with your horse that will make it hard not to smile. Since you made the effort to acquaint your horse with the whip in the longeing lessons, this is also the time you can begin to carry it in your liberty exercises. As in previous Conversations, it can serve as a useful extension of your pointer finger as you work to talk to your horse's Buttons.

*11.4 A & B*
*Gretchen and Image practice following a target hand and Matching Steps (A), followed by Drop It to Stop It (B).*

## Conversation: Moving on Target

❶ Standing parallel to your horse's shoulder in Dancer's Arms with your knuckles held near his muzzle as a target hand, ask him to walk on.

❷ Reach out one of your feet in a large step forward as you point at your horse's Girth Button to urge him to join you in Matching Steps. Include verbal cues such as a kissing noise or saying, "Walk." Drop It to Stop It and reward often (figs. 11.4 A & B).

❸ Then turn your torso slightly as you are moving together, remembering to hold your leading hand out in front as a target for the horse to follow. Use the angle of your belly button and Core Energy to create an arc or to negotiate a change of direction.

*11.5 Pressing the Follow Me Button tells the horse to come with you as you walk forward.*

❹ Another way to engage your horse in the targeting Conversation is to press your leading palm against the upper neck about four inches behind the horse's ear on the Follow Me Button (fig. 11.5). Then place that hand out in front of him as the target. This is what his mom did to tell him, "Come along, now." Not all horses are receptive to this, though some will follow you right away.

❺ Entice your horse to check in with your target hand after any stop.

**6** As you evolve your target hand, you can easily switch to walking normally at your horse's side as though you have an imaginary lead rope and are reviewing leading and Matching Steps from earlier Conversations. You can use the Mid-Neck Button as needed, and even add obstacles like we did in Step 5 on p. 90 (figs. 11.6 A–C).

**7** From playing with your target hand, Matching Steps, and Dancer's Arms, you can step farther away from the horse and more into the center of the pen. Asking a horse to move out and away from you can be interpreted by the horse as displeasure. You may have to Send your horse out and away from you and Beckon him back several times to ensure he understands that you are still having a Conversation—it's now just long-distance. When your horse is clearly confused about leaving you, aim the whip toward his Go Away Face Button and then the Shoulder Button. Since you have already built trust and respect with these buttons, he should stay calm when you ask.

This level of liberty Conversation in the round pen can be a little tricky because your angle of movement—that is, where your eyes and belly button aim—might actually tell your horse to stop, change direction, slow down, or speed up. If you, or your Core Energy, accidently gets ahead of the horse's girth line, for example, you will slow your horse, change his direction, or even stop him. It can be helpful to occasionally hold the whip up to your belly button so you can see where it is aiming. This is a very good check when you repeatedly get a troubling response or odd answer from your horse, even though you are sure you are asking the right question.

If you have trouble achieving connection with your horse at a distance, go back to his side and do tandem

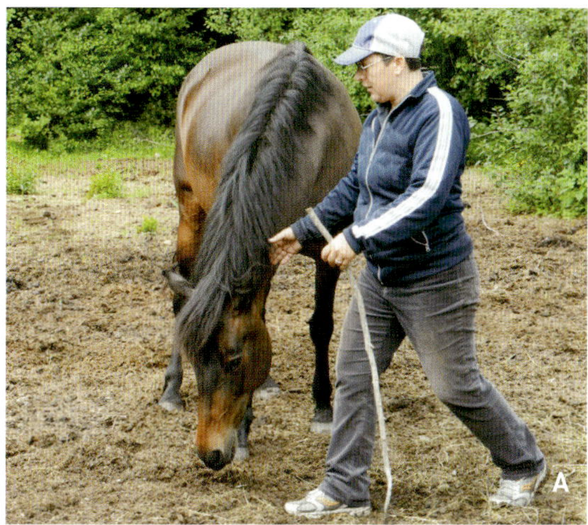
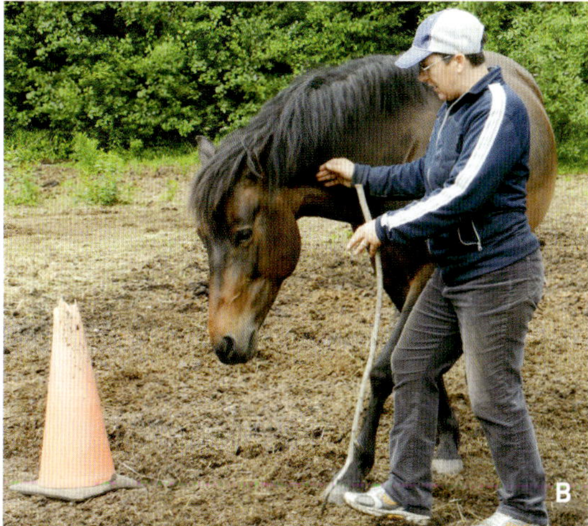
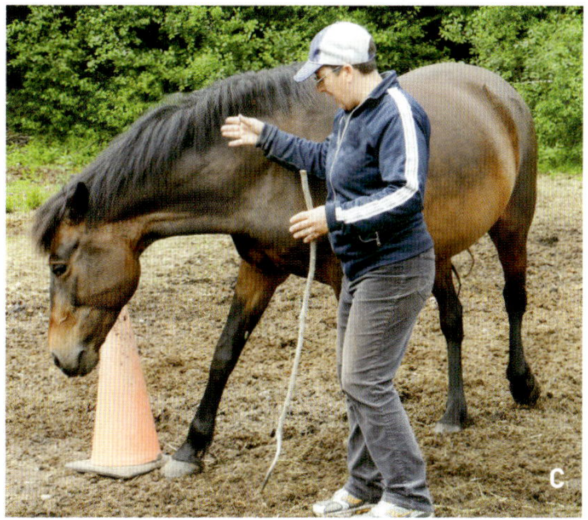

***11.6 A–C*** *When I use the Mid-Neck Button on Luna, she does Aw-Shucks (A). We move forward toward an obstacle added to the pen for interest—the cone provides a visual destination (B). I Stomp to a Stop and Luna "kicks" the cone, just for fun (C)!*

Matching Steps, gradually moving away from him as far as you can while keeping the connection you've reestablished. Exaggerate your steps and Drop It to Stop It often to offer a reward. If your horse comes to you, simply use his Buttons to gently Send him back to his own circle, and repeat the Conversation of "starts" and "stops" at least three times. Now he will begin to believe there won't be any "chasing," even as you move to apply pressure to his Girth and Hip-Drive Buttons.

## Picking Up Speed

When you get to the point where your horse will walk on his own arc away from you for a full half or complete circle while responding to your questions and calmly stopping, you can open the Conversation up to trotting. This requires the same actions as when you did Trotwork from the outside of the pen (see p. 166).

### *Conversation:* Trotting on the Inside

❶ Hold your Dancer's Arms extended outward for the trot. Your target hand in front must tell him *forward* and not to come across in front of your space. For clarity, I recommend pointing the finger of your target hand. Essentially, this hand shows direction, while the other hand controls the whip, speed control, and distance between you (fig. 11.7 A).

❷ Low-step jogging in place might be the best trot cue for your horse. Make sure you are either moving parallel to his front legs, or if facing him on the circle, your belly button should line up with his Girth Button. When you drop behind his Girth Button you change the game of going faster to one of chasing.

*11.7 A & B*
*Gretchen uses Dancer's Arms and Matching Steps at a low jog to initiate a trot (A). You only need one or two steps at first. Stopping right away prevents the horse from feeling chased and increases trust between you (B).*

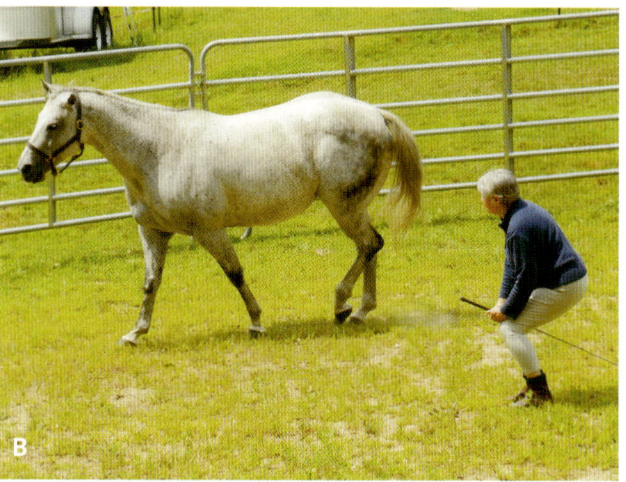

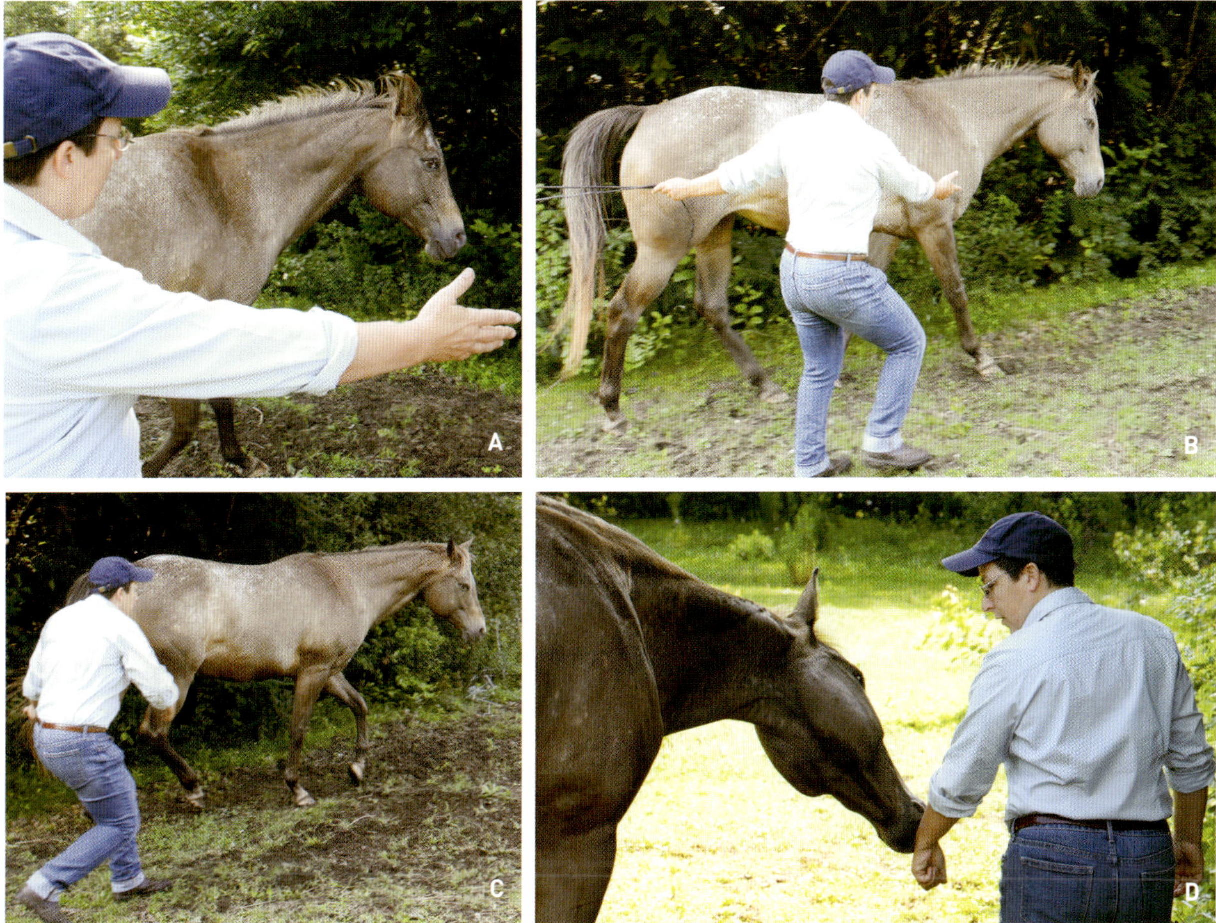

**3** Some horses are content to have you apply pressure at the Girth Button as a cue to pick up speed. If needed, point the whip and your eyes toward the Hip-Drive Button to ask for an upward transition.

**4** All you need is a couple steps of trot at first and then Drop It to Stop It and praise your horse (fig. 11.7 C).

**5** When your horse easily picks up a relaxed trot and stops on cue, you can ask him to change direction. Step across to the other side of the pen while the horse is out on his circle. Open into an "X" and block the horse's circle. Your Core Energy will cause him to to turn around and arc the other way.

**6** When you get three changes of direction, ask him to turn to face you. You can do this two ways: using your target hand to Turn the Canoe (figs. 11.8 A–D) or going to the middle of the pen and making a big "O," perhaps even dropping the

*11.8 A–D Dancer's Arms tell Dakota to stay inside the "frame" I am asking for, and on the arc I'm gesturing to, at the trot (A). By drawing my target hand back toward me, I ask Dakota's head to begin to turn back to me (B). When I Drop It to Stop It and bring my target hand into "O," Dakota begins to Turn the Canoe to face me (C). We check in for a moment of praise (D).*

*11.9 A–C  I purposefully step across Vati's path, aiming my Core Energy in front of her (A). When I turn my belly button abruptly and make an "O" with Drop It to Stop It, she halts sharply (B). I increase my "O" to include my neck and shoulders, taking the pressure off and saying, "Thank you" (C).*

whip (figs. 11.9 A–C). Look down or look at the Hip-Drive Button, asking him to, "Move over to the side." Your horse will Turn the Canoe and look at you, which should bring him to a full stop.

**7**  If your horse is too excited or too intense after trotting to approach you, try pointing the whip at his chest, in front of his body, and holding your other hand up the way a policeman does to stop traffic. You can do this even if you are on the side of the horse. Drop the intensity in the pen by raising one hand up high, blowing out a big breath, and then dropping the same hand low. Drop the whip if necessary. You are signaling that he should lower his intensity and his head. Note that if you ever feel your horse gets overexcited and you cannot keep your Inner Zero, you must safely leave the pen and regroup.

In a short time your trot circles will feel like magic and your horse will enjoy showing you his beautiful motion—perhaps even showing off by arching his neck and looking really proud, inviting you to call out to him that he is definitely the most stunning horse in the world (fig. 11.10).

## Conversation: Go Ahead, Canter!

The next logical step up from trot is canter. If you are having wonderful and fulfilling Conversations about all the things you can do together at liberty, your horse may offer (on his own) to canter one day. If not, it is reasonable for you to ask him to try.

*11.10 By pulling my belly button in, and bringing my target hand in, as well, I ask Dakota to tuck her belly up, drop her croup, and arch her neck, resulting in a lovely trot moment.*

**❶** While he is in a nice, forward trot, "scoop" the whip up from the ground toward his Hip-Drive Button in one, big swoop. You may instantly find yourself skipping into a canter as you Match Steps into the movement you desire (the old "canter movement" you used as a child when riding an imaginary pony works well). Adjust your intensity level to your horse's needs. If he acknowledges your request in *any* way—his ears become concentrated; he slightly speeds up; he lifts the head—quit your cues. You asked for something, you got something, so say thank you.

**❷** Next time you ask, see if he needs more engagement from your Core Energy. If he attempts a stride or two, wonderful, but do not let him continue. Ask for a stop and reward with his favorite scratching and perhaps Rock the Baby on his withers. Again, the reason to stop him is that you do not want him to begin to run away from your cues or feel pushed. By allowing him to gradually acquire confidence at the canter, you will gain both trust and respect. The canter should feel like he has accepted your request from a place of interest, curiosity, and joy. Progress is made when you learn to calibrate your own movements with clarity and the right level of intensity. This way you are telling him you are pleased with his effort.

**11.11 A & B**
*Liberty can turn into a wonderful game (A)! When I "sit" and spread my arms in front of Dakota, I am rewarded with a little impromptu rear (B).*

## AN OPEN FORUM

Obviously, you may not be able to build up to the canter at liberty all in one day. When things start to get confusing for either of you, cease all cues, go to your horse, breathe, scratch, Cup his neck (see p. 92), reward him, and call it a day. Every Conversation builds on the previous one, so the next day so you can run through all the movements again several times and arrive back to where you were confused, most likely bringing more clarity to your own movement. Do not have strict agendas or hard-and-fast goals. Liberty Conversations are meant to be a lot of fun, so when you stop enjoying it and it doesn't feel playful, it is time to quit for the day (figs. 11.11 A & B).

Liberty Conversations are two-sided events. You are not just showing up and expecting him to conform or perform your agenda. You are opening up a forum for sharing and mutual admiration. This type of relationship will foster the horse's desire to bond with you, and when he does, he will look out for you the way a best friend does. You both will have definite likes and dislikes, and up-and-down days, good ones and bad ones, like anyone else. Through the medium of Horse Speak, however, you will be aware of these shifts and either ask him, "What's up?" or just acknowledge that you "get it," which in many ways, for a horse and a human, is enough.

Some horses want you to push all their movement Buttons from the ground, yielding sideways and backward and playing target games before they are mentally ready for riding. Some horses do not seem to need as much pre-riding discussion. Certain horses have a huge sense of humor and will try to make you laugh. Other horses don't want to talk that much and are content just Sharing Space in their Bubble. The miracle of Horse Speak is that now you can actually begin to know *what* your horse really wants and how you can best nurture a partnership with him.

## AT LIBERTY

I FIRST LEARNED ABOUT the round pen through the work of Monty Roberts in the early 2000s. Roberts had worked out a system based on understanding that certain movements the person makes in the round pen triggered the horse to respond in a particular way. The end result was something he called Join-Up®, in which the horse seemingly chose to come to the person of his own accord. Roberts' intention in bringing this to the public was to facilitate people's understanding that horses are in fact listening and waiting for us to give them the right signals so that they understand they can and should trust us.

Since I first studied Roberts' methods, round-penning has been used by many others who feel they have a bead on the techniques and how to use them best. I have been to many such clinics, observed numerous trainers' tactics, and heard all manner of opinions on the round pen and its use. Some trainers show great finesse; others use round-penning like a system of punishment. I actually heard one trainer state that you should not consider your horse "joined-up" until you can get 200 "inside turns" without argument. (An inside turn means the horse changes direction by facing you first, as opposed to a defensive change, in which he turns his hind end toward you.)

Other sayings I recall from round-pen clinics include: "Take his breath away from him, and only give him a chance to breathe when he is thinking the way you want him to," and "If he isn't sweating, you aren't making your point." In more generic terms, there is an idea associated with round-penning—and with longeing, too—that the horse needs to "get his bucks out" by running in circles.

In my work with horses, I have had my hands on hundreds of them: rescues, as well as school, show, and therapy horses. In my experience, I have learned that horses *do not* round-pen each other in the way so many people today round-pen their training projects. They do occasionally run around together, blowing off steam in the fresh spring air, and enjoying a good romp. In addition, I have watched horses school, assist, and educate each other in enclosed spaces. For example, when one horse (Horse A) discovers another (Horse B) is in distress, Horse A will establish he is the teacher, guide, and therapist for Horse B. Horse A will drive Horse B forward a tiny bit—just enough to get Horse B to move off softly, without resistance or worry. Secondly, Horse A will usually exaggerate his own footfalls, making clear, slow, heavy steps, often with a lowered head. He usually Scans the Horizon and maybe blows a Sentry Breath. He may perform the Nurturing Breath and Beckon Horse B toward him to do the Greeting Ritual over and over again. Depending on what Horse B's problems are, Horse A will figure out what Horse B needs the most, and then he will work on that.

When it comes to the round pen, I need to be Horse A with Joe. Joe is in the round pen now. He is staring at me with wary eyes. His owners have tried to "join up" with him, and I appreciate their efforts to try and help Joe feel better about his world. Unfortunately, he mentally and emotionally shuts down when he is put on a circle—whether longeing or round-penning. To him, the circle is the place where people chase him with a whip and he has to pass a test he never understands before they finally let him stop and put a saddle on instead. My first approach in this space must relieve him of his belief that I am about to chase him around.

> It is difficult for people to phase their intensity in a clear, precise way.

To do this, I will stay outside the round pen. By staying on the outside, I remove a whole level of pressure. Joe is already tense: His eyes are wide; his body is "taller" as he prepares to run; his breath is shallow, and his head is high. I begin by establishing the level of "X" that I will use, plus the level of intensity that goes with it (see p. 100). Horses are the most confused by our lack of clarity in regards to stages of pressure. Of all the things that I spend the most time teaching students, it is the levels of intensity that cause the Number One disconnect between what people intend and the actual message they transmit. To review:

■ *Level One* intensity includes "looking at" the horse. A horse is constantly aware of other horses' "eyeball pressure"—for example, when the lead mare looks at him, it means she may want his hay pile, and it is best if he gets off it before she gets closer. Noticing another horse's "looks" saves quite a lot of scuffle.

■ After a clear stare-down comes a "head snake" that looks like the horse is flattening his ears, and nosing definitively in the direction of another. This is *Level Two*. Level One meant, "I intend to claim your hay pile," while Level Two is a request for movement, "I am coming toward your hay pile, and you have a full minute to start leaving, please."

■ *Level Three m*eans business—and the action, intensity, and movement reflect this: "I am at your hay pile, you are not moving away fast enough, so I am opening my mouth and preparing to bite you!"

■ *Level Four* insists in no uncertain terms: "I am standing on your hay pile, you are slow as molasses, and now I will take a chunk out of your hide!" Then, as we learned in Step 7 when I first discussed levels of intensity, when the dust settles, everyone returns to Zero as though nothing ever happened.

It is difficult for people to phase their intensity in a clear, precise way, and then drop all the way back to Zero at the drop of a hat. It takes lots of practice.

Joe sees me look at him and reads that I have an intention (Level One pressure), because he holds his breath and stares fixedly at me. I acknowledge his "right answer" by stepping backward and looking down in Aw-Shucks. We engage in this version of Approach and Retreat three more times. Each time he sees me look at him and answers by looking back at me, I take the mental pressure off by doing Aw-Shucks. This process invites him to finally take a huge, deep breath, and do an Aw-Shucks himself. I am overjoyed that he can ask me this question—he is asking if it is all right to relax. I answer by Scanning the Horizon dramatically and then blowing a Sentry Breath. He does Copycat, which means that he would very much like to follow my lead.

I step up to the fenceline and make a dramatic "O," then hold out my knuckles for a Greeting. He wastes no time, striding over to check in, but he does do one more Aw-Shucks on his way. This is to let me know he means no pressure against me and wants me to feel comfortable with his Approach. I am very pleased with his efforts. I tell him what an impressive horse he is—so brave and trying so hard to do the right thing! I tell his owners also that we have achieved a kind of "join-up."

Of course, Mike and Liz are confused by this since what Joe and I are doing looks very different from the other kind of round-penning with much running and sweating. They are relieved that Joe wants so badly to be a "good horse," and that our connection has happened so fast and with so little effort!

The day is not over, however. Joe has to learn to trust me and listen to me, even when my intensity phases up. If I end our session now, it will not serve him entirely because I need him to feel comfortable with *all* my levels, just like he would with another horse. The only way I can do this is to work through the levels carefully, faithfully returning to Zero each time. It is not the pressure that scares the horse, it is the lack of returning to normal that upsets them.

I am in the moment with Joe, and he can relax because I am going to prove to him that I can go to Zero, whatever he does. I go through all the stages of the Greeting Ritual to reassure him that nothing has changed between us. When we get to the part where I ask him to Yield Space to me in the front, he turns and walks away. Fantastic! I only asked for a step backward, but him giving me the whole side of the round pen is a nice compliment.

Because it is not what I asked for, I can see that Joe overcompensates for what he *thinks* people want and isn't really listening to what they are *actually asking for*. Now I know that he guesses and "overgives." Either way, he has set me up perfectly for the Sending position, so I move along the fenceline into his Hip-Drive Button space.

Aiming my whip up toward the Hip-Drive Button of his hip, I keep my intensity between a Level One and Two. I move toward him around the outside of the round pen, and my movements are deliberate and steady: each footfall is slow and anchored to the ground. Joe can perceive me coming and is watching. He knows I am pointing at his Hip-Drive Button, and at first, he seems to resist my approach. Joe then lays his ears back in defensive annoyance, nods his head, swishes his tail, and moves off too fast.

Joe is saying that he feels defensive about being driven forward by me (or all people). He is unsure why I am Sending him, because he gave me the whole fenceline, and why do I need to move him again? He doubts I understand the Hip-Drive Button, and the Tail Swish says he doesn't like this. I cannot gain his acceptance until he knows that I know how to drive.

Because he moved off (even though it was too fast) I step backward and do Aw-Shucks. This is the best way to tell him in his words that he did a good thing and that I appreciate it. He trots to the other side of the round pen, so now I approach his Hip-Drive Button again, with the same slow, deliberate steps. This time he leaves sooner and swishes his tail more dramatically. He is telling me he thinks I am picking on him. He cannot differentiate my driving him from behind from bullying.

Again, I do a brief Aw-Shucks. He looks back at me hopefully, and I nod my head, Scan the Horizon, and do a Nurturing Breath. I really want him to start understanding that I am not going to act like a rabid dog but like a mother horse. This time, when I start to move toward his Hip-Drive Button, he kicks out, squeals, and runs off, shaking his head. This means he really feels defensive, picked on, and that life is "so unfair right now." I have not gone above a Level Two intensity. Now that

> I wait, breathe, and pull at a little grass to demonstrate my dedication to Inner and Outer Zero.

Joe is feeling picked on, I must respond with Level Four so he can tell the difference. I have watched countless horses do the same thing: They make their upper stages of pressure clear and obvious so that young or immature horses can understand what qualifies as the low end of the spectrum.

I take one definite step away from the fence and Joe, raise the whip in one, fluid motion, and crack it down swiftly onto the earth. Then, I completely stop.

Joe scoots forward, then turns and faces me with a questioning look. I step back and sit down on the ground, combining Aw-Shucks, Sharing Space, and Zero. I sit like this until I see Joe shift in a small way. I am undoing all his confusion about pressure, Buttons, movement, and whips. Joe needs time to think because I just showed him a big Level Four intensity, followed by a dramatic and certain Zero. Before, he consistently overreacted to my Levels One and Two so he needed the contrast to understand what I was doing was definitely *not* Level Four. And he needed to see how committed I am to going right back to Zero. Now he stares at me, breathing steadily, almost without blinking. I wait, breathe, and pull at a little grass to demonstrate my dedication to Inner and Outer Zero.

Joe lowers his head slightly and closes his eyes. Joe's owners are nearby in chairs, watching, so I explain everything that has just happened. I say his sleeping is not ignoring me at all; he is absorbing the Conversation. Mike and Liz admit that Joe has never slept in the round pen—ever. (Once in a clinic, I actually had a sensitive, nervous horse lie down

## AT LIBERTY

*I begin connecting with Joe from outside the round pen, using my "O" Posture to Beckon him over (A). Inside the pen, I use Dancer's Arms and Core Energy aimed toward the Girth Button as one form of pressure to ask him to move (B). Soon we move off together, Matching Steps (C).*

and go to sleep while in the round pen, with his buddy standing watch.)

After about four minutes (which can seem like forever when you are just sitting there) Joe stirs a little. I stand up and ask him to move one, single foot backward by aiming my whip toward his toe and staring intently at it, initiating Fun with Feet. He pauses for only a moment, then carefully, moves that one foot backward. Eureka! I have attained the beginning of healing! I back up, blow out a deep breath, and say, "Good job." I then ask for the other front foot to move backward, and he moves it almost instantly. I stop, breathe, and tell Joe what an amazing horse he is and that I am proud of him.

For years I have noticed when we begin to reach through the "fog" and communicate with a horse using Horse Speak, he then seems to understand the intention we send with our human words infinitely better. At times, it almost seems as though the horse has some rudimentary grasp of our words.

Either way, I enjoy using my voice to convey my praise and respect for a horse and model this for my students.

I step backward, make a defined "O," and invite Joe to come toward me. He Pauses for a moment, then, Licking and Chewing, he nods his head and steps toward me all the way up to the fence. I have just Beckoned Joe, and he has Come Back. We Share Space for a moment, enjoying our newfound connection.

Now I must go inside the round pen and do everything all over again to make sure that this connection will be honored even though the pressure will be heightened when I step to the other side of the fence. On my entrance, Joe stiffens and lifts his head high. I Scan the Horizon and blow the bogeyman away, adding a Nurturing Breath as I soften my whole being. I want him to read my level of intensity—currently Zero. If he reacts to my Zero as though it is a Level Three or even Four, things could get out of hand very quickly. Greeting him with a Knuckle Touch and

Cupping his neck, I repeat the Fun with Feet game we did when I was outside the round pen. He reacts to my Level One intensity (simply looking at his front foot) by moving his hoof back one step. He now realizes that I am discreetly Sending him without any pressure directly on him. Making an "O," I invite him over. He takes one step toward me and freezes, which is fine. I let him feel "stuck" for a brief period before I casually glide over to him to do a Knuckle Touch. As we move to Copycat, he takes a Shuddering Breath: two breaths in and one long breath out. I use this moment to reach up and Groom his withers, adding Rock the Baby. I sway slightly, holding my whip in the Walking Stick position and humming under my breath. Joe adjusts his front feet to a wider stance, and I can feel the tension oozing out of him and into the ground.

> **He is assuming we have arrived at the point where I chase him.**

After several minutes, I step back, moving my eyes, belly button, and whip tip toward his Hip-Drive Button. Joe politely steps away with his hind end then sails off quietly toward the edge of the round pen. I nod my head and do Aw-Shucks as a "Yes." Gently, I aim my eyes again at his Hip-Drive Button. With a blow of his nose and a shake of his ears, Joe marches off in a quiet, slow walk. I step back and praise him. When he stops, I turn away and Scan the Horizon. Every time I Scan the Horizon or blow a Sentry Breath, I am acting like a concerned leader who will watch out for predators and protect Joe.

Now, I make Dancer's Arms and aim my Core Energy at his Girth Button, and as he walks off, I turn and casually join him, Matching Steps with his stride and stepping in alignment with his shoulder. We are Going Somewhere Together. When I stop my feet and drop my pelvis in Drop It to Stop It, Joe stops on a dime. I want to keep Joe focused on my intention and my levels of intensity so that he is actually listening to what I am saying and not anticipating, so I repeat the pressure on the Girth and Hip-Drive Buttons, and we move one step at a time in Matching Steps.

Joe needs to stay present and keep talking with me in order for me to give him the leg-up he needs in his life. We get to a place where he moves away nicely at a walk, changes direction softly, halts when I halt, and even walks off shoulder to shoulder with me and follows my direction, so it is time to ask for a little speed. Asking for speed is tricky with panicky horses. They assume that if you are asking them to run, there must be a reason for it.

I step to the center of the round pen and ask Joe to move off at a walk, which he does very politely and with a sideways ear, floppy lip, and low-head attitude. Carefully, I lift my knees a little higher, and nod my head to signal I want a speed change. I am stepping in time with his front feet in the center of the ring, and I use the whip so slightly—still between a Level One and Level Two intensity—to indicate a little more energy.

Joe lifts his head, tightens his jaw, and pins his ears, accompanied by a Tail Swish. I am not surprised he feels pushed or bullied by my request: he is assuming we have arrived

at the part where I chase him. I continue to talk to him, telling him that I understand his worry. Keeping my cues securely at Level Two, I make sure my inner energy is Zero. This sounds like an inner-versus-outer conflict, but keep in mind that horses truly never like to leave Zero on the inside. Our goal, too, is to be able to stay Zero inside, regardless of our cues and gestures outside.

I want Joe to experience the feeling of his own Inner Zero. I want him to be able to find an inner calm while his body transitions to a faster gait. I have to model this of course, so I exaggerate my "O" Posture and deepen my inner calm as I also ask him to politely take a step or two into a trot. He shakes his head in protest while stepping up into a defensive trot. The moment he does this, I Drop It to Stop It, looking at the dirt, and blowing out a sigh of relief. It still takes Joe several steps to notice.

I let Joe stand and look at me for a full two minutes. He is breathing steadily and flicking his ears. His eyes are blinking, and he looks as though he is sorting out some big thoughts in his head. When he starts to soften just a little, it lets me know we can try again. I ask him to walk off, which he does nicely, then right away I ask for one step of trot. I use all my body language and even my human words, saying to him, "Just one step of trot, please." Joe walks faster and faster, head up, ears in a concentrated backward hold. His chin is a bit

tight, but his eye and closest ear keep flicking onto me. This tells me he is trying to "get it."

He lifts into a hesitant trot. Immediately, I Drop It to Stop It and add a human expression to tell him how wonderful that was: I clap my hands. Clapping hands is a normal impulse for us; it expresses joy. Horses fling their tails with dramatic flair when they are very pleased with themselves, and I want Joe to understand this similar connection to joy. He is a little startled by the noise of the clapping at first, but he soon senses my happiness and gives a contented tail swing, "blowing his nose" in response.

I tell his owners that the next time he trots they too are to clap and tell him how wonderful he is. Once again, I ask Joe to move off, and this time he has soft eyes, listening ears, and lightness in his step. As soon as I nod my head to indicate a speed change and point at the Hip-Drive Button, he wastes no time in picking up a light, airy trot. I jog along in place next to him *at* the Girth Button (not behind him) in time with his feet for about five steps. Then I Drop It to Stop It, and he practically slides to a stop, too.

We all clap and shout out words of praise and admiration for his wonderful, soft, and lovely trotting! Joe "blows his nose" again, "wags" his tail, shakes his head, and grunts. All this says he is proud of himself, and he enjoys the attention and praise!

# Now You're Up There... What Should You Do?

## MOUNTED CONVERSATIONS

I wanted to know, without a doubt, whether my mare Dakota wanted to be ridden. I needed to hear her, within the context of her body language, actually say to me, "Hey, I've got this great idea. Do you want to get on me and we will Go Somewhere Together?" I could only hear her answer if I knew her language. My search for the answer to this question launched years of study, resulting in the book you are now reading.

While loose in the riding ring, Dakota positioned herself at the mounting block and looked at me. Her invitation was the answer I'd been searching for. Since then, I do not ride a horse that does not willingly come to and peacefully stand at the mounting block; one that does not welcome me onto his back.

In this chapter, I offer Conversations that clarify your body language while mounted. All our Conversations pose questions and allow you to observe your horse's replies. They are not meant to train or teach the horse anything new or different. They are meant to increase your proficiency in Horse Speak, while sitting on your horse's back.

### Around the Mounting Block

How do you think *your* horse feels about being mounted? Horse Speak can make mounting a positive experience for you both. It is easy to unbalance your horse when you mount him, and you can also unbalance him when you

## Keys to
## Horse Speak: Step 12

dismount. Learning to take your time so you can both stay at Zero in the process of mounting and dismounting helps everybody stay balanced and neutral.

First, *really notice* how your horse reacts to being mounted. (Consider asking someone to take a photo of your horse's face while you get on. You can even do "before" Horse Speak and "after" Horse Speak, and compare them.) A stoic horse may grimace while being mounted. A sensitive horse may raise his head and show anxiety. An energetic horse moves off when you step into the stirrup. There are many possible reactions. When looking at your horse, notice his ears, eyes, and in particular, his mouth. What you have long thought was acceptance, may instead have been be acquiescence.

## Core Concerns

Your Core Energy can cause confusion when mounting, especially with a sensitive horse. Say, you lead the horse to the mounting block with Dancer's Arms, then step up on the block. Your Dancer's Arms helped the horse know where his head should be as you approached the block, but, when you take your target hand away and turn to face the saddle, you may put Sending pressure from your belly button onto the horse. He will naturally swing his head toward you and his body away, in response to the Sending message your body is conveying.

To clarify your body language, practice mounting with your Core Energy turned toward the horse's head. You can also diffuse your horse's anxiety about mounting with the following Conversation.

## *Conversation:* Mounting with Breath

❶  Begin by leading your horse to the mounting block and position him as if you are going to mount, but instead just sit on the block for a few minutes (Retreat) and breathe with him. Breathe long enough to see your horse visibly relax next to the block. This is a good exercise some evening when you don't have time to ride but do want to have a Conversation with your horse. Tack up in your normal routine and have a *Breath Conversation* at the mounting block. Try to sync your breath to his. Observe the subtle language he shows. Take really deep breaths; try a Shuddering Breath (see p. 26).

❷  The next time you walk to the mounting block, do Matching Steps and look to the horizon in two opposite directions. This tells your horse you've checked for prey and are taking care of him. If your horse keeps looking off in the distance, go to

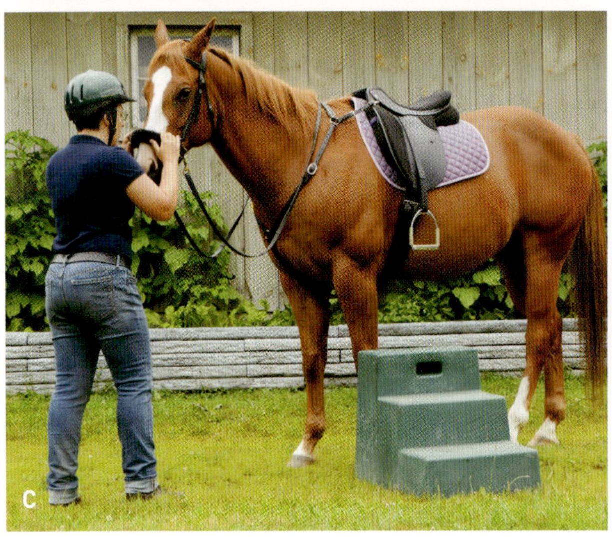

**12.1 A–E** *Gretchen Matches Steps with Clark on the ground, on our way to the mounting block (A). He tends to be a tense horse, so connecting with him and asking him how he feels before someone gets on will help communication with him in the saddle. I use Sentry Breath to "blow away the bogeyman" (B). When we reach the mounting block, I Rock the Baby on his bridle (C) and withers (D) before doing Approach and Retreat with my foot and weight in the stirrup (E).*

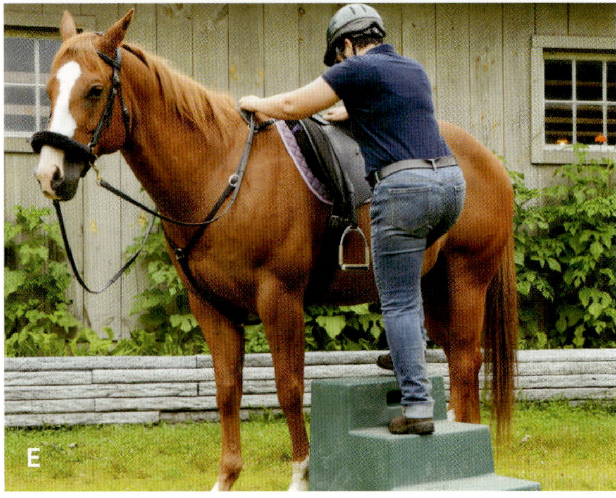

Level One intensity and blow Sentry Breath in the direction he is looking toward. The horse understands you are blowing away all his bogeymen (figs. 12.1 A & B). After Sentry Breath, exhale deeply, look at the ground—exaggerating your "O" Posture—and do Aw-Shucks. Phase up and down in levels of intensity from Sentry to Aw-Shucks as often as you need to until your horse drops his head. For some horses, this Conversation can set a positive tone for the entire ride.

**3** Show your horse affection before you mount. Before getting up on the mounting block, check in with a Knuckle Touch and sniff your horse's neck (see p. 122). Reach up and lightly scratch the Friendly Button where the forelock meets the forehead. Most horses also appreciate having each front foot picked up and moved in a gentle circle at the mounting block—it releases tension.

**4** Rock the Baby first on his bridle while standing in front of him, and then while standing on the mounting block with your horse in position in front of you, facing the same direction as your horse with your hand closest to him on his withers (figs. 12.1 C & D). Shift your weight from one foot to the other or from one hip to the other. Remember to sync your rocking to your breath, and breathe as slowly and deeply as you can. Your horse may take a step to rebalance himself. Many horses are taught to stand still no matter how awkward and unbalanced they feel. Letting him widen his stance may be a huge relief to him. Also some horses appreciate Rock the Baby at the mounting block with one hand on the withers and one behind the saddle.

**5** Now, once you mount, dismount again immediately, and walk your horse in a medium-size circle. Bring him back to the block, breathe, and mount again. Repeat this sequence three times, paying attention to your horse's comfort and body language. If there is any tension stop, breathe with your horse, do Aw-Shucks, and then resume the Conversation.

**6** Try a Copycat Conversation with your horse about the mounting block. Lean over him slightly as if preparing to mount, and then lean back upright or away from the horse. Repeat, syncing your leaning toward and away from the horse to your own breathing. Do this at least three times before getting on and staying on. When you repeat this Copycat every time you mount, at some point your horse may simply lean toward you as you step in the stirrup. What a wonderful way to start a ride!

*12.2 A & B Once I am on Clark, I sit quietly and breathe for a moment (A). When we are in the saddle, our legs wrap around the horse's lungs, so breath becomes a significant means of communication. I use one hand to gently Rock the Baby at the withers as I focus on deep breathing (B).*

## Letting Go

Controlling the horse's movement is what we consider riding. Before you begin to dictate your horse's movement, reap the rewards of sharing breath, balance, and unity to build new levels of trust and respect for both of you. Allow your skeleton to move with his skeleton.

When you are on your horse, your legs are wrapped around his lungs (fig. 12.2 A). Think about that. Your breath sends important messages to the horse via your seat and legs when you ride. In a perfect world, all riders would become addicted to taking big deep breaths when they are on horseback. If possible, try the following Breath Conversation bareback or with a bareback pad. If using a saddle and you feel safe, consider dropping your stirrups. Having a friend on the ground with a lead rope the first few times you explore this Conversation can help you both relax.

### *Conversation:* Gumby Pose

❶ Mount using the mounting block techniques on p. 187. Reach down and Greet your horse, then rest your hands, palm down, on either side of your horse's withers (figs. 12.3 A–C). Hold a rein in each hand, but don't make contact with the horse's mouth. Don't worry about the slumped position this puts you in—like a limp Gumby. Try and feel your horse's breath. Elongate your inhale and exhale. See if you can sync your breath with his. Perhaps he will sync with yours. Just sit there and breathe with your horse.

**2**  Observe any messages of discomfort or tension from your horse. Even though when you get on, you habitually want to do what you always do, I ask you to suspend what you know about riding for a little while. Instead, observe what happens when you try and listen to your horse. You both have ideas about what happens after you mount. For example, mares, in particular, want to do things "right," and they want *you* to do things "right," as well. This is *not* business as usual. An incredibly simple Conversation may take some time to absorb.

**3**  Try Nurturing Breath in the Gumby Pose, making a soft inward snorting sound (see p. 25). As in all your Conversations, you are saying something to the horse. If you lean forward and hug his neck, take the time to hear what he says in reply (fig. 12.3 D). There aren't right or wrong answers.

**4**  Have your friend lead your horse off at a walk if you have a groundperson. Or take a few steps forward if on your own and feeling secure. Keep breathing and stay in Gumby Pose (fig. 12.3 E). When you are relaxed, you can't help but feel how the

*12.3 A–E  When I am up on Rocky's back, I reach out to Greet him. This gives us a chance to feel connected like we do on the ground (A & B). You can try reaching out to check in. The Gumby Pose helps me become anchored to Rocky (C). I lean forward to hug his neck, similar to the way I invite horses to "hug" me on the ground (D and see p. 121). Then in Gumby, I breathe and encourage Rocky to walk a few steps forward (E).*

horse's shoulders swing forward with each step. You are not trying to influence the horse's body. The idea is to do Matching Steps: Follow the horse's movements with your own body. Just like when you were Matching Steps on the ground, you almost can't help but have *that same Conversation*, now with your knees in step with the shoulders of the horse. Let each knee rock forward when the shoulder on the same side moves. Keep breathing and Matching Steps for a few minutes as long as it is comfortable for you both. Your body is absorbing how it feels to connect breath and movement in a physical Conversation of unity.

**5** In Gumby Pose, keep your hands on the horse's withers and continue to allow your knees to follow the movement of his shoulders. Without stirrups to brace against, you'll notice your hips sway in rhythm with the horse's walk. When you bring your awareness to your hips, you will realize you are having the Rock the Baby Conversation with your hips. Or, perhaps your horse is rocking *you*.

## A New Greeting

Mounting is a Greeting, and being mounted is Sharing Space. I've developed a *Rein Conversation* that serves as a check-in and Greeting from the back of the horse. It simply explores how your horse feels and thinks about the movement of the reins. You're just saying, "Hello," to your horse.

### Conversation: "Hello" Rein

**1** To begin saying, "Hello," do some breathing in Gumby Pose then softly sit up. By now your horse shouldn't be confused about standing still after you mount. Have a friend hold him if he needs in-hand support the first few times you try this.

**2** At Zero intensity, hold the reins palm down—not completely slack but without contact. Lift one arm up about 12 inches from your lap with your palm still facing down (figs. 12.4 A & B).

**❸** Breathe in while you lift and breathe out when you lower your hand, three times. I recommend you say, "Hello," out loud with each lift to reinforce your subtle body language.

**❹** Try this with each arm. If your horse moves or shifts, just allow it until he is ready to stand still again. If he is fussy, walk or have your friend lead you in a large circle, then return to a halt. Resume the rein lifts. Most horses eventually drop their heads and take big, sighing breaths.

**❺** Don't forget to praise him and blow some long out-breaths as a reply and reward.

Horses are often afraid to get the answer wrong, and you are doing something different in the saddle by not asking for anything. He may be stressed because he doesn't know what you are asking, or by now he may be curious. Or, he may not respond at all. All you are saying to your horse is, "I'm doing this (lifting the reins). What do you think about it?" You are also letting him know you're not going to get angry if he doesn't understand something. He may brace until the twentieth time you lift a rein, then all of a sudden, drop his head and give you the big sigh.

It is a new page for you both. Instead of manipulating the horse with the reins the instant you mount, you are waiting to see how he responds to the fact that they are there and you are using them. Your horse gets to realize he is not in trouble if he responds with curiosity to this rein movement, and you have created a way to Greet him with "Hello" Rein.

This is a Conversation of mutual receptivity. Can you do something with the reins you've never done before? Can you move the reins without having any agenda?

### *Conversation:* **Copycat Reins**

Now we ask the horse if he wants to move his head by softening his neck. It is the same Go Away Face Yield we did when we practiced "hugging" on the ground (p. 121). Although your horse may think you want him to move his feet, you are not asking him for this at this time. A friend standing in support you from the ground can be helpful.

**❶** Start with your hands on the reins, palms down, without making contact with the horse's mouth. Do a "Hello" Rein, then lift one rein at a 45-five degree angle from your body. Let your gaze follow the back of your knuckles.

*12.5 A & B When you ask for Copycat Rein you say the same thing you said with it on the ground: "Will you follow me?" With Copycat Rein, I simply ask Clark's head and neck to soften (A), and Rocky's, too (B).*

**2** Your arm and shoulder make a sweeping, open arc to the side, which encourages your horse's head and neck to arc, too.

**3** Allow the horse about 10 seconds to reply before you bring your hand back to the starting position. Breathe in while you lift, and return the rein with the subsequent out-breath. All you are asking for is for him to look in the same direction as your hand and eyes have traveled. He may do so right away in Copycat or he may move his head in some other way. Explore three outward lifts on each side. Reward him and praise, no matter what he does.

So many horses are confused about where their heads should be in ground-work and even more worried about where they should be when you ride them. Horses communicate with each other via head gestures with each other. We effectively make them mute when we force them into exacting positions with harsh bits, over-checks, and tie-downs. With this diagonal lifting and lowering, all you are saying is, "Please relax your head and neck," and, "Remember to look in the direction of the rein." In addition, the Conversation releases tension in *your* neck and shoulders.

## Conversation: **Up-and-Over Rein**

The first two Rein Conversations show your horse something really different is going on now. This third Conversation is the mounted version of Move Your Feet Over, which you had on the ground (p. 64).

On the ground you've been routinely asking your horse to yield his head away from you. You touched the Go Away Face and Mid-Neck or Shoulder Buttons to ask for his front feet to step over. When mounted, you use your foot pressure to cue him to follow Copycat Rein into a "move-your-front-feet-over" cue.

**❶** Have a rein in each hand with extremely light contact. Do this without stirrups if you feel secure, or have a friend support you with a lead rope from the ground.

**❷** Use the rein on one side as you did in Copycat Rein, but this time you follow the motion with your eyes and open your knee on the side of the rein cue, taking its weight off the side of the horse and adding a cue to move, such as heel pressure, a kissing noise or saying, "Walk."

**❸** The goal is to have him take one step over with his one front foot. Your horse may simply lean a bit or shift his weight to the other shoulder. Praise any try, go to Zero, and breathe. You are creating a Conversation that causes you both to become very clear about the body language that connects you both to your horse's front feet. Try this on each side, coordinating hand, eye, and knee gestures at the halt, before you try it at the walk.

*12.6 A & B  I use the Up-and-Over Rein to ask Clark and Rocky to Go Somewhere (A & B).*

*12.6 C & D Yielding the front end gains respect under saddle, just as it does on the ground. This Rein Conversation helps Rocky and me get in step at the walk (C), and relaxes his neck and shoulders, totally phasing down intensity (D).*

**4** Next, ask your horse to walk on, engaging the Up-and-Over Rein in either direction, giving the horse an "opening" sensation as you lift your knee on that side slightly off the saddle or bareback pad. Use the same flow of Dancer's Arms to open up for the turn as you did with the longeing and liberty Conversations on the ground. Your horse will flow into the turn as your weight automatically shifts to the hip outside the bend (figs. 12.6 C & D). Try this in each direction to gain more unity in your timing, just as you did while Matching Steps on the ground. Both you and your horse will feel more confident as a result.

The Up-and-Over Rein should feel like you are broadening your chest and opening up your arms—again, just like you are doing Dancer's Arms in the saddle. This Rein Conversation not only keeps the connection to your horse calm and relaxed, it automatically balances your seat because you cannot open your arms without sinking your Core Energy downward.

## Building Body Awareness

So far you've had the mounted Conversations with the palms of your hands facing downward. Before trying the next Conversation, explore how using your hands palms or "fingernails" up changes your torso. The position of your hand influences the rest of your body language.

## *Exercise:* Palms Down vs. Fingernails Up

❶  Sit on the edge of a chair that has a firm seat. Pull your belly button in just until your chin naturally tucks somewhat and your lower back straightens. You might notice your pelvis is now tilted backward, putting more pressure on your seat bones.

❷  Engage the muscles of your core slightly and hold both arms out from your body with your palms facing the floor. Softly curl your fingers as if your hands are lightly holding reins.

❸  Lower and lift your extended arms, one at a time at a 45-degree and a 90-degree angle from the front of your body. Note the sensation of the change of emphasis in your seat bones, core muscles, and breathing.

Now repeat the exercises, facing your palms or "fingernails" up. Again, pay attention to how your torso and seat bones shift, and how your breath feels. You may have to breathe in while you lift your arms with palms up. Feel which muscles are now engaged around your ribs and body's core.

Because of the weight shift in your torso, using your hands *palm down* influences the forehand of the horse and using them *fingernails up* affects his hindquarters. Now you are ready for the next Rein Conversation.

## *Conversation:* Fingernails-Up Rein

This Rein Conversation sends your balance through your rotated arm and shoulder blades, over and down to your opposite hip. This occurs because when you rotate your arm and hand to turn your fingernails up, your body engages the *opposite* leg in a natural, diagonal vortex. This allows the horse to pivot with ease, because you have created a Pivot Point inside yourself. You'll explore Conversations with your horse's haunches with your hands in this position. This is essentially the same rotation of the arm you used in the Therapy Back-Up (see p. 81).

❶  If possible, do this Conversation without stirrups. Ask a friend to help on the ground if it helps you feel more comfortable.

❷  Beginning at Zero with your hands holding the reins in normal riding position at the halt, rotate one hand at the wrist until the palm is up and you can see your fingernails (fig. 12.7 A).

**3** Engage the rein by bringing it toward your belly button, just until you feel the "ripple" of the horse beginning to move (fig. 12.7 B). You should *not* pull all the way to your belly button—instead, release at the first sense of movement in the horse and take a big breath. You might only come back a quarter of an inch before you get a twitch or shift from your horse in his back and hip on the same side as the rein. Or you might get a response as soon as you turn your fingernails up. You are beginning to "talk" to his hip.

**4** Try the other hand. When you turn up your fingernails and draw your hand toward your belly button, see how sensitive you can be to any subtle shift the horse makes. Feel a sinking in your seat bone as your horse releases his hip slightly. If your horse is sensitive, he may step under with his hind leg on the side you engage the Fingernails-Up Rein (fig. 12.7 C). If your horse gets confused by this Conversation, go back to previous Conversations and review.

**5** Now try the Fingernails-Up Rein at a walk. Again, have a friend lead your horse if you need to. Follow the movement of your horse's front legs with your knees, keeping your palms down, feeling and breathing with the movement of your horse.

**6** For best results, begin with the rein closest to a fence or boundary so your horse doesn't

*12.7 A–C I rotate my arm so I am holding the rein fingernails up (A). Drawing the rein toward my belly button, I feel for the shift of weight in my seat, as well as in Rocky's back (B). Rocky steps under his body with his left hind leg in reply to the Fingernails-Up Rein Conversation (C).*

think your cue means "turn." Rotate your arm, fingernails up. Hold the rein in this position for a few strides. You will begin to feel your horse laterally bend under you. Release and praise him. This strengthens his core, back, and hips, one side at a time. It is like yoga for the horse!

**7** There is a rhythm to the hand movements in this Conversation. When you are ready to stop, it helps you both if you have a chosen point in the arena or a cone set to mark where you are going to begin your halt. Your body will send subtle shifting messages that will help each of you prepare, but to halt from this Conversation, you are going to engage one rein at a time so you can feel the horse shift his weight to his haunches. Simply use the Fingernails-Up Rein for a stride, then rotate your other arm into Fingernails-Up, as well.

**8** Say, "Whoa," sink in your seat bones, and put just enough pressure into the reins to halt. This mirrors the Drop It to Stop It language you use on the ground because when both your hands are in Fingernails-Up Rein, your pelvis "sits down."

## Claiming Space in Front

Your Fingernails-Up Rein Conversation prepares you for the Therapy Back-Up under saddle. In the back-up, you ask the horse to yield the space in front of his body. You are still claiming the space in front of your horse, even though you are mounted. He may have reasons why he thinks yielding his escape route to you is a bad idea. The more you reward every flinch or lean backward with a quick go-to-Zero response, the more he will be motivated to back up at the most subtle of cues.

### *Conversation:* Mounted Therapy Back-Up

**1** Begin this Conversation without stirrups if you can, and perhaps with a friend holding your horse. Breathe in and rotate both arms so you can see your fingernails. Feel how this influences your body as your belly button automatically comes back toward your spine, shifting your weight. Your horse's hips are influenced by this shift in your Core Energy.

**2** Breathe out. At the same time, draw your hands back toward you, just until you feel a flinch or a lean backward

*12.8 A & B Clark tries Mounted Therapy Back-Up for the first time (A). One step is fine! I release and praise him (B).*

*12.8 C & D Rocky prepares to back up as I rotate my arms (C). When my heel touches his Jump-Up Button, he lifts his belly more and really sinks his hips to take a step back (D).*

from your horse (figs. 12.8 A & B). Reward his response. If there is any resistance or bracing, adjust your intensity level. Try moving your feet slightly forward at the Girth Button, as though your calves were squeezing toothpaste out of the tube. Or a slight touch of the heel at the Jump-Up Button can get a reply (figs. 12.8 C & D). When your horse yields the space in front of him to any degree, drop to Zero, breathe, and praise him.

All of these Conversations are intended to clarify your body language while mounted on your horse. The "masters" at the great riding schools were not mastering horses, they were mastering their own body language so their communication while riding the horse was crystal clear. Most horses love to explore our ideas and may even have some of their own. They are curious to see what we'll come up with next—let's make it something rewarding and fulfilling for them, as well as us (fig. 12.9).

*12.9 Who knows where our Horse Speak Conversations will take us next?*

# UNDER SADDLE

WHEN I NEXT see Joe, a few weeks have gone by since our round-pen Conversation. Mike and Liz are overjoyed with the changes they sense, not only in Joe, but in themselves. They have each worked with Joe loose in the round pen, starting from the outside, then cautiously expanding their Conversations with him to the point where he will not only trot in an airy, comfortable, and calm manner, but has even lifted up into a gentle canter on two occasions. His frantic mad dashes are history.

Mike and Liz did not even realize how tense things really were between them and Joe until they changed. There is a new sense of acceptance of each other's moods. Now, if any one of them is in a funk, they make new choices based on the depth of the quality of their relationship.

I had asked Joe's owners to not ride him until their relationship was sorted out. Having agreed to wait a short few weeks, they have come to realize how much they truly love this horse. Now I stand back and watch Mike and Liz approach, halter, lead, and round-pen Joe. I say very little—a few pointers here and there—but I am overjoyed by their progress. I think back to the nervous and confused relationship they all had endured until recently.

Liz confesses that while she feels calmer around Joe than ever before, she still carries some old tension and fear from the memories of Joe's past explosive moments. She says she can feel herself lose her Inner Zero at times. But Mike is ready to ride Joe. We put Joe on the cross-ties. I tell them not to go blind, deaf, and dumb while they are doing routine tasks around him. I remind them to stay present and remember that Joe still has his Bubble of Personal Space—but on cross-ties, he is unable to defend it.

Quickly, they notice that his normal "dancing around" in the aisle is actually a constant attempt to deal with the accidental Sending messages their body language produces. By aiming their eyes, belly buttons, and hands all directly at Joe, they are in a constant version of "X." If they simply aim their torsos slightly *away* from Joe while they are grooming or tacking up, they avoid Sending him with their Core Energy. When they have to face Joe, they adopt more of an "O" Posture, drooping their shoulders slightly, breathing deeply, and softening their overall intensity. Joe visibly relaxes. Mike resumes brushing while maintaining a slight 45-degree angle to Joe in his "O" Posture.

I show Mike how to hold the saddle while facing Joe's rump, swinging it lightly up and over his back, making a neat arch with Mike's own posture so that he ends with his right arm hugging the saddle on Joe's back and his belly button facing Joe's head. This little "pirouette" is so easy to learn, and takes all the rigidity out of saddling. I learned it when I worked on a dude ranch and had to swing a dozen, heavy Western saddles onto horses all day. It not only saves *your* back, but the horse's, as well, because it causes the saddle to touch down lightly and gently.

Mike says Joe is not good at being girthed. I suggest he lightly bump the horse's girth area with the girth as rapidly as possible for about a minute, *without* buckling the girth on. This is a way to get Joe used to the sensation without completing the action. It tells Joe that we respect his feelings about this procedure. After a full minute, Mike lets the girth drop and scratches Joe's favorite spot, while telling him how good he is. Joe has visibly relaxed, so now I take the girth and release a deep sigh as I lift it and connect it to the saddle. Joe makes a slight face as I tighten the girth, so I release the girth again and step back. Now Mike does the same procedure. We are not using some special technique to "train" Joe to be girthed up, rather we are using the depth of our new relationship with him to convey a different idea about the procedure. Joe may *never* like being girthed, and that is his prerogative. He only has to understand we mean no harm by it.

Once Joe has his bridle on, we make our way to the round pen. Even though Mike and Liz have a small riding ring, all their new work with Joe has been in the round pen. To help set them up for success, we'll stay in a place where they have all been practicing being calm with each other. I remind everyone to breathe and maintain Inner Zero.

I lead Joe up to the mounting block, then past it, while paying strict attention to his expressions, thoughts, and feelings about it. I can see apprehension, tension, and also the effort he is putting into trying to be a "good boy." We walk up to and past the mounting block three times before I climb up and step down, followed by stepping up to the saddle while on the block and touching it with one hand, three times in a row. Then I place my belly over the saddle, step down from the block, and walk away, and repeat three more times. This may seem agonizingly slow, but the effort pays off. Joe now appears clear, focused, and relaxed.

Whenever Joe feels stressed, he reaches for my knuckles, reassuring himself by sniffing my hand for a moment. I know that he needs this check-in to steady himself, and so I offer a Knuckle Touch regularly. He seems unsure where his head should be, so as I stand on the mounting block, I stabilize his bridle the way I have stabilized his halter on the ground. "Holding hands" with his mouth on the reins with one hand, I Rock the Baby with my other hand on his withers. At first, he does not understand; he nods his head and wiggles a bit. I am not worried about his confusion, telling him with my voice that I just want him to relax and find his balance. Mounting a horse knocks him off balance, which Joe hates. Slowly, he relaxes into my reins and takes a step to the side with one front foot, widening his stance, and making himself more grounded.

I sigh and he copies me, "blowing his nose" and letting out a Yawn to boot. I reward his Yawn, getting off the block and walking him in a small circle. We repeat Rock the

## UNDER SADDLE

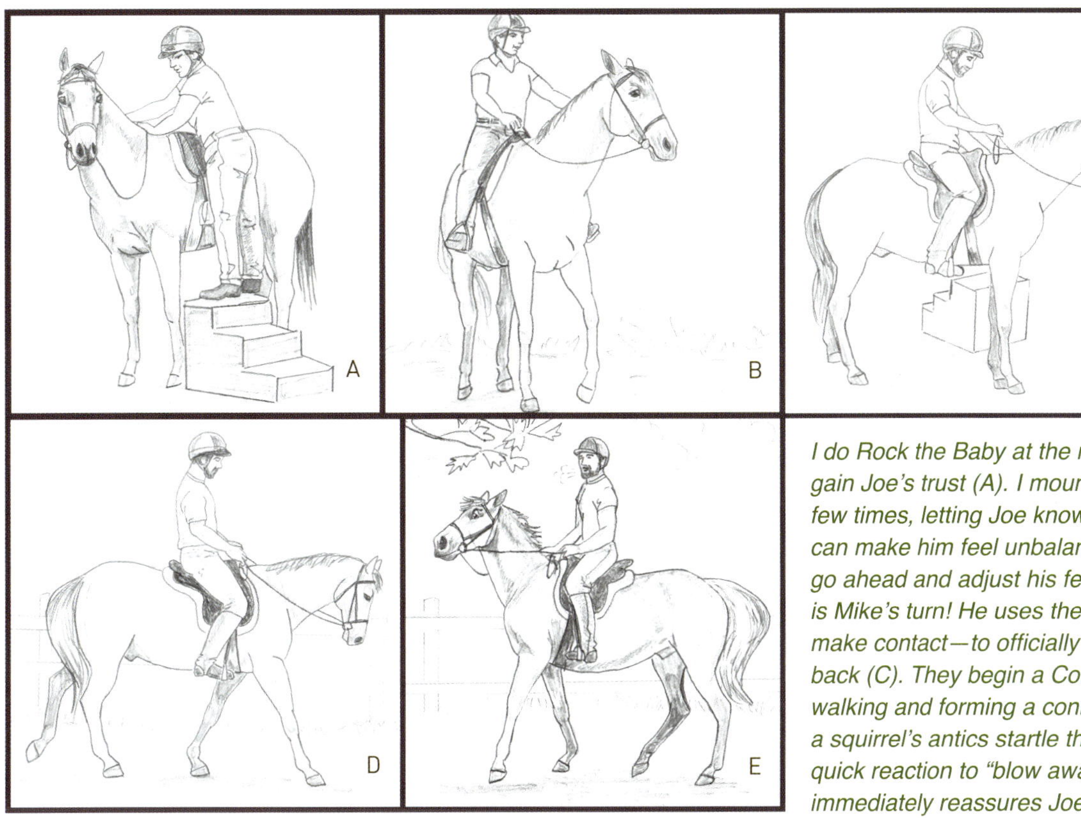

*I do Rock the Baby at the mounting block to gain Joe's trust (A). I mount and dismount a few times, letting Joe know I understand it can make him feel unbalanced and he should go ahead and adjust his feet (B). Then it is Mike's turn! He uses the "Hello" Rein to make contact—to officially Greet Joe from his back (C). They begin a Conversation about walking and forming a connection (D) when a squirrel's antics startle them (E). Mike's quick reaction to "blow away the bogeyman" immediately reassures Joe.*

Baby two more times, and on the third time, Joe steps lightly to the block, holds his face in a comfortable position, widens his stance, then leans visibly into my leg; he is telling me to go ahead and get on.

I gently scoot into the saddle but do not put my feet in the stirrups. As soon as I settle into the saddle, I Rock the Baby on his withers for a moment, then dismount slowly and carefully. Dismounting also makes horses lose their balance. I want Joe to be completely comfortable with every single part of these procedures. Too often, horses just endure these everyday activities. After mounting and

dismounting two more times, I am ready to hand Joe over to Mike who does a wonderful job repeating all the steps. Even though I just went through every facet of mounting with Joe, Mike has never done this before. Joe needs to become familiar with Mike's "accent" every step of the way as they continue to learn to communicate. Mike ignores Joe's muzzle seeking his knuckles a few times, and Joe gets upset, swishing his tail. Mike appears to be about to correct his horse, so I step in and explain that Mike missed a cue from Joe. Mike chuckles and not only offers to check in with Joe with a Knuckle Touch, he also announces he sees that, going forward, Joe is going to help him learn what to do.

As soon as a loving Greeting is complete, Joe literally *looks* at the mounting block! He is engaged and ready to go forward. Mike is amazed at what he is witnessing, and Liz can't help but laugh, saying that she never had a clue how intelligent horses are and she wishes she had learned to "talk" to them *years* ago.

Joe and Mike march up to the mounting block, Matching Steps. Because Joe has initiated the trip to the block, we move right into steadying his face with the reins and Rock the Baby on his withers. It takes Mike a minute to figure out how to Rock the Baby gently. At first, he nearly knocks Joe over, who reprimands him with flattened ears and a tail swish. But Joe does not step away from the mounting block. He is now confident enough in his relationship with Mike to give his owner feedback.

The two step away from the mounting block and repeat this cycle for two more rounds. On the third round, Joe leans into Mike's leg, who is overwhelmed by the idea that his horse is *inviting* him to mount. Mike has never felt anything like this before. He dutifully mounts and dismounts three times, and I give him pointers about softening his movements a little to help Joe keep his balance. Mike admits he never thought about *how* he moves could disrupt Joe's balance so much. I explain it is like supporting and carrying an uneven backpack.

> Mike admits he never thought about *how* he moves could disrupt Joe's balance so much.

Mike is in the saddle and he and Joe are quietly connected and enjoying each other's company, when a squirrel makes a poorly judged leap and lands heavily on a branch just outside the round pen. It causes all of us to jump a little. Mike automatically blows Sentry Breath at the sudden sound, and even though Joe is startled, he does not take a single misstep. The horse makes a visible effort to return to Zero. He licks his lips, "blows his nose," waggles his ears, and swishes his tail at the tree. Mike reaches down to Rock the Baby on his horse's withers, and laughs out loud at the event. He and Liz marvel at their "new" Joe. The "old" Joe would have taken off bucking, for sure. I remind them that this is the *same* Joe, but now we are all his herd-mates, and he trusts us.

I ask Mike to just sit still in Gumby Pose and breathe together with Joe for a moment. We are ready to say, "Hello," with the reins. I have discovered that many horses may know *some* things about reins but are guessing when it comes to others. It is best to review what reins are and what they mean in bite-size chunks, just like we did with the mounting process.

First off, I have Mike simply lift a rein straight up in the "Hello" Rein Greeting Ritual. Joe stands there and doesn't fuss. After three "Hellos," on the right rein, Mike does three on the left. Joe blows through his nose and licks his lips. He appears relieved to be Greeted in this way. Mike now pulls one rein over to the side, doing the Up-and-Over Rein, encouraging Joe to yield his head to the side—three times each way. Mike follows the rein motion with his eyes, looking in the direction of his own hand.

Joe feels Mike's head change direction, which tells him Mike is leading, and he is following Mike's lead. Mike Grooms Joe's withers with Rock the Baby between each cue.

Mike now lifts one rein, draws it to the side while turning and looking in that direction, and gives a "kissing" noise to let Joe know he wants him to not only move his head but take his feet with him. Joe carefully crosses his front feet and walks a small circle. I tell Mike to quit asking, and return the rein to Zero, saying, "Whoa." Joe stops moving immediately. As we repeat this, Joe becomes more confident and responsive with each time.

Persuading a horse to yield the space in front of him is an essential part of herd hierarchy, and we have found several ways to do it on the ground. It is now time to show Joe that Mike "owns" the space in front, even when he is in the saddle. This is essential to helping Joe learn not run off with his rider. If Mike "owns" the space that Joe is walking into, then every step will be "given" over to Mike, avoiding the tug of war many people have with their horses.

Fingernails-Up Rein is next, and it aligns the bones of Mike's arm and scapula, anchoring his Core Energy and balance in his pelvis. This sends a clear and resistance-free signal to the horse's hips to yield backward, and Joe does so flawlessly when Mike gently pulls back on the rein. One step is all that is needed—almost immediately Mike is able to have Joe "roll" his hips backward with almost no effort. I tell him to refine the arm motion down to the subtlest cue. We hardly see Mike do anything while Joe lifts and rounds and steps deeply under himself. Both horse and rider are very pleased with themselves; Mike says he never knew he could have such a good time going backward.

I ask Mike to dismount for a minute so we can let everything soak in. Forgetting himself, he hops off in the vaulting-styled leap he is used to. Joe pins his ears and steps sideways. Mike realizes without my saying anything that Joe found that surprising and uncomfortable, saying he will be more mindful in the future. Liz chimes in that it is going to take a lot of work for them to stay present to the fact that much of what they consider "normal" behavior around horses is actually worrisome to Joe.

From now on, they will be able to explain to Joe about the ins and outs of what they want to do while sitting on his back. Their homework is to set up cones and barrels and to spend the week simply walking in circles and figure-eights. By practicing Up-and-Over and Fingernails-Up Reins, Joe will continue to relax, soften his neck, and use his hips with more balance and rhythm.

They should practice the Therapy Back-Up on the ground, as well as in the saddle to strengthen his core. In addition, halting is, in fact, a shortened back-up. If Joe can balance over his hips every time he halts, he will not push against the bridle. I think the day will soon come when Mike feels like the trot will be as airy and natural under saddle as the one they've cultivated in the round pen. He and Liz are reaching for a new relationship with Joe with ever-increasing appreciation for how safe they all feel with each other. Joe needed to feel safe with Mike and Liz, just as much as they needed to feel safe with him. Through Horse Speak, their friendship is now developing aspects of *trust, respect, unity,* and *acceptance.*

# *The White Mare*

Crawling upon the back of a white Arabian mare, I toddled through horsemanship.
On her back, I discovered I could cling and fly at the same time.
There, I only knew timelessness—*unum est omnia*.

At the end of my marriage, a white mare came to me in a dream.
As I scrambled onto her sleek back, I knew my life would speed up now.

Years later, I had another dream; post-apocalyptic this time.
A group of women formed a giant statue of a white mare placed upon a wheeled cart.
She was to be paraded through the town as a symbol to put down weapons, and join together.
I arrived just in time to form her head.
I woke with the sensation of my hands dripping in plaster as we affixed the mare's head
    to the statue.
Only then, did I ever hear of the ancient Goddesses: Epona, Macha, Rhiannon, Demeter.
I never knew a king had to be married on the back of a white mare
    to symbolize his devotion to the land.

One by one, the white mares marched into my life, brought forth by the women
    who owned them: Nefertiti, Sugar, Trix, Bella, and finally, Karma.
Each mare was a Goddess in her own way. Each mare wounded in her soul.
Of the hundreds of horses who crossed my path, I began to take notice
    that the white mares signified something more.
Each white mare took me around a different bend.
The last mare—Karma—led me to my co-author, Gretchen. Karma now lives with my herd.
I have never sat upon her tired back, but in her rebirth here, she was given a new name:
    Parvati, the Goddess who brought the teachings of divinity here to Earth.
Vati is here, reforming her head-space, and has come to a deeply peaceful place,
    leaving her private wars behind.
She led me into the world of writing.

Who is saving whom?
In a post-apocalyptic world, it only matters that people come together and find renewal.
May the apocalypse be only a personal destruction of what is useless, old and haggard.
May the parade of people rally behind the White Mare, and find not only a new love for her,
    but also for themselves.
May the blessings of the White Mare bring renewal to our world.

—Sharon Wilsie

# ACKNOWLEDGMENTS

Thank you to my amazing partner, Laura, who has never faltered in her support and enthusiasm, and to my co-author, Gretchen Vogel, without whom this book would never have gotten off the ground.

Thank you to Trafalgar Square Books for your belief in this book.

Thank you to my bevy of friends and my loving family for all your love and kindness.

Most of all, thank you to all the horses....This book is truly dedicated to you.

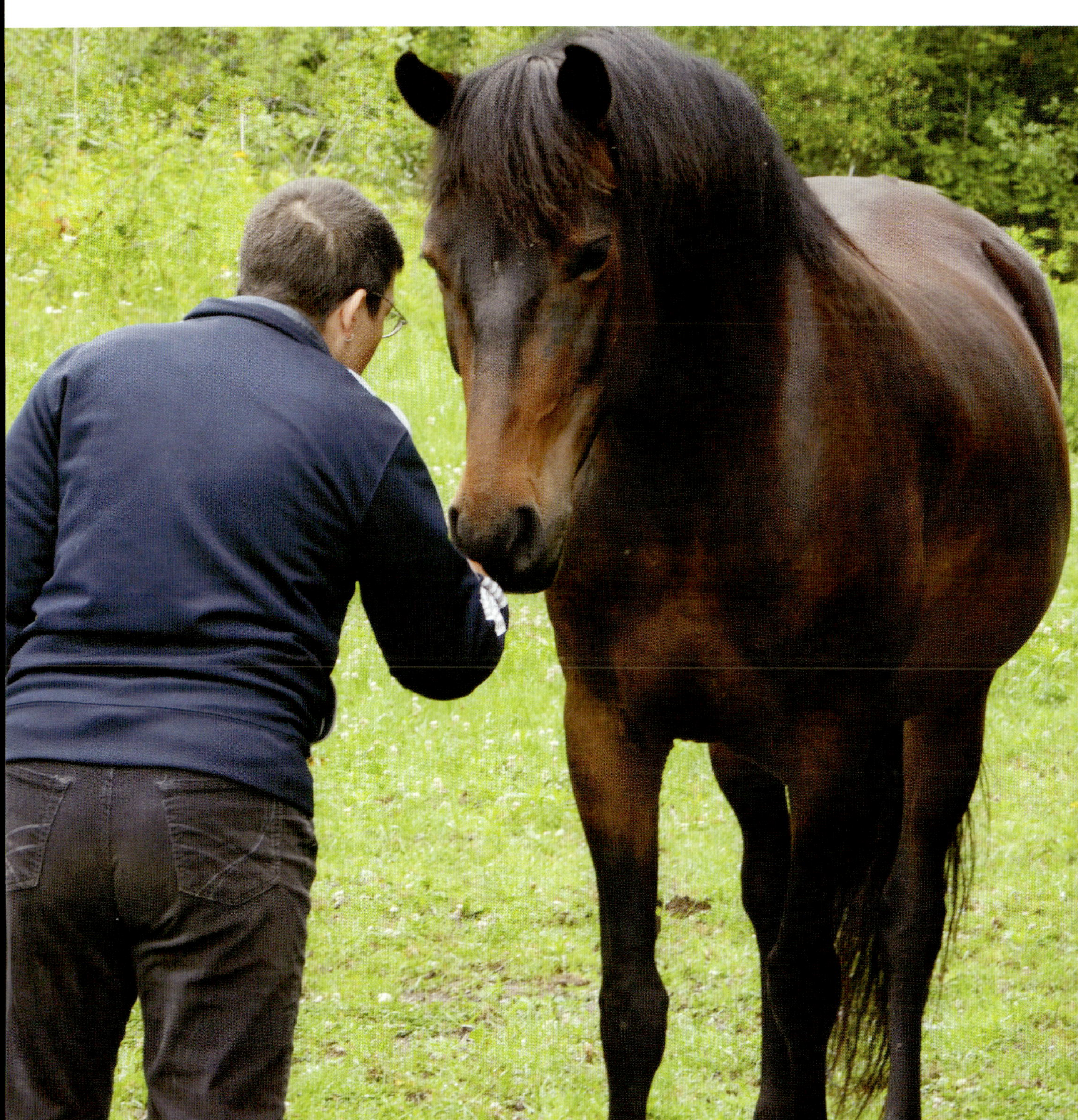